Next Generation Network Services

Next Generation Network Services

Technologies and Strategies

Neill Wilkinson
Quortex Consultants Ltd., UK

JOHN WILEY & SONS, LTD

Other Wiley Editorial Offices

John Wiley & Sons, Inc., 605 Third Avenue,
New York, NY 10158-0012, USA

WILEY-VCH Verlag GmbH
Pappelallee 3, D-69469 Weinheim, Germany

John Wiley & Sons Australia, Ltd, 33 Park Road, Milton,
Queensland 4064, Australia

John Wiley & Sons (Canada) Ltd, 22 Worcester Road
Rexdale, Ontario, M9W 1L1, Canada

John Wiley & Sons (Asia) Pte Ltd, 2 Clementi Loop #02-01,
Jin Xing Distripark, Singapore 129809

British Library Cataloguing in Publication Data
A catalogue record for this book is available from the British Library

ISBN 0 471 48667 1

Typeset in 10/12pt Palatino by Deerpark Publishing Services, Shannon, Ireland

Printed and bound by CPI Antony Rowe, Eastbourne

Dedications

I dedicate this book to my deceased parents, Irene and Bill, who unfortunately will not see it published, but supported me throughout my childhood and through to the early part of my degree. They always told me I would do something valuable and I'd like to think this is it!

I'd also like to dedicate the book to my wife, Catherine, and son, Thomas, for supporting me in this project. They have both given me motivation to complete this task.

And finally to my cat Callie who spent many a fond hour on my lap asleep when I couldn't sleep because of the work involved in compiling the book.

Dedications

Contents

Preface

The mother of invention

Telecommunications is now the fastest changing part of the IT industry, encompassing vast disciplines from distributed systems to real-time applications. I have had the pleasure of being involved in what I believe is the most exciting time in its history. It wasn't always this way, as telecommunications started out as a novelty: *"An amazing invention, but who would want to use one?"* – US President Rutherford B. Hayes after making a telephone call from Washington D.C. to Philadelphia. Interestingly however, in 1879, the first telephone was installed in the White House. At first it was hardly used, because there weren't many other phones in Washington to call [WHITEH].

From the now immortal words 'Watson, come here!' Alexander Graham Bell's humble telephone[1] has become the most ubiquitous device on the planet. Followed swiftly in 1891 by the invention of Almon B. Strowger's patented system of automatic switching. The Strowger switch design was so fundamental that it soon became the backbone of the World's telecommunications network for at least the next 100 years. The story has it that Mr. Strowger (at the time an undertaker in Kansas City) was so incensed by a competing undertaker getting business and not him, because a cousin of the competitor's worked as an operator. He decided to remove the need for operators![2] From the first 99-line automatic exchange installed at La Porte, Indiana, in 1892, the telecommunications industry has never looked back. The UK followed suit and in 1912 the first experimental automatic exchange was installed at Epsom, by the then Auto-

---

[1] Alexander Graham Bell patented the telephone 14 February 1876.
[2] How true this story is I don't know, but I like it anyway.

matic Telephone Manufacturing Company (later to become Plessey, then GPT and now Marconi). The rest as they say is history!

The invention of the transistor in the late 1940s also had a profound effect on telecommunications; the eventual demise of the electromechanical Strowger exchange, to be replaced by the current day Stored Program Controller (SPC) exchanges. Over the last 30 years the telecommunications industry has been gradually improving the SPC exchanges with more features, better software engineering techniques, and increasing their capacity with more and more powerful processors. Most recently, the use of *standard* computing hardware and the influence of the Internet have lead to the so-called next-generation networks. It is these next-generation networks and the new capabilities and services that this book is about.

Motivation for the book

My aim for the book was to bring together all the technologies that at first glance seem unconnected. My view is that nothing in telecommunications is unconnected – it is the nature of the beast. One unique event can change the whole focus of the business. In recent years, that event has been the global acceptance of the Internet as an acceptable form of remote communication (read telecommunication). This development has turned conventional telecommunications on its head. The bursting of the Internet market bubble in 2000 has also caused a ripple across the telecommunications industries, with what effect, only time will tell. My own belief tells me there will be a renaissance in telecommunications – the phoenix from the ashes if you like.

I read many technical books in my role as a consultant. I need to know the answers to questions my clients ask me. My annual book budget is more than I would like to admit and certainly more than my wife would like me to spend! So I wanted to avoid the problem that I face, by putting a reference together that would cover at a reasonable level all the areas that influence and will influence telecommunications services and networks of the next generation. What I cannot achieve is a complete reference, nor would that be appropriate. I would have loved to be able to include all the technical specifications, standards and other references that I refer to in my day-to-day role. That would have resulted in another Tolstoy and I wanted the readers to enjoy the book and be happy to follow up the reference material if they wanted or needed to know more.

I wanted to stimulate interest and thought in the telecommunications community about the massive task we are about to undertake. That of swapping out the whole of our current voice telecommunications networks from a circuit switched solution to a packet switched solution. We are in a golden age of telecommunications, never before has it moved so quickly or changed so dynamically (I saw at a recent conference it likened to the US gold rush). We need to seize the opportunity of the convergence of voice networks and the Internet to deliver exciting and useful services.

Who should read this book

The smug answer is of course everyone! More realistically, the intended audience is telecommunications professionals who are working on current circuit-switched networks and are looking to see how the technologies they are working with will change over the next five to ten years.

Equally, engineers who are now facing the new data to carry (voice) should find this book a useful reference to show where the voice networks have come from and are going, and how this influences their role.

How to read this book

The book is split into three main parts: technology, services and implications, which tries to sum up where it is leading. If you are coming from a data networking background, then you can skip the technology section pertaining to packet switching and IP in the first part of the book. If you are from a telecoms background, then you can probably skip the section on circuit-switching techniques in the first part of the book.

Everyone should find section two on services interesting. The service examples I use come from a mainly telecommunications service focus; these are the kind of services I am closest to.

Section three is where I attempt to apply some perspective to how long the services may take to develop, how money might be made from these services and who will implement them, and best of all who might use them.

The topic area is replete with acronyms, so as a mechanism to improve readability I have included at the end of the book, a list of acronyms and a brief explanation of each. In addition more information about the areas covered in the book is available at the web site http://www.telecomsoap-box.org.uk. This site contains white papers and urls relating to topics covered in the book. The author can also be contacted via this web site for comments and questions about the book.

And finally I hope you enjoy reading the book as much as I have enjoyed compiling and writing it. Special thanks go to my publisher for allowing the book to be written and showing faith in me, and specifically Sally Mortimore for getting it off the ground and Birgit Gruber for keeping it going. Thanks also to Quortex Consultants for giving me a stimulating environment to work in. Don't let anyone tell you writing a book is easy!

Part I: Technology

INTRODUCTION

As technology advances, it is a standard axiom that new ways to exploit that improvement are found. Take fibre optic cables, the first cables were used to carry hundreds of telephone conversations by utilising a single wavelength (colour) of light and modulating it with a time division signal. The creation of Dense Wave Division Multiplex (DWDM) has made that same fibre optic cable capable of carrying a significantly larger volume of information, combining many time division signals together on to the same fibre by utilising different colours of light.

Faster processors have enabled software engineers to create more useful tools and languages such as Java. Increased processing power has allowed the execution of real-time applications without having to resort to code optimisation techniques or machine language coding. The improvement in tools and programming languages, combined with the greater processing power, has lead to the proliferation of more complex services.

The massive acceptance of the Internet and better tools has meant anyone can become a web-savvy individual with their own home page. This will directly influence the capability of users of telecommunications service to manipulate and tailor their service to their own personal requirements.

This new empowered consumer will be able to use tools and facilities provided by the service providers to control, even provision new services. This will have to happen as the complexity and number of users of services increases, the cost to a service provider to provision and maintain these services will increase. Customer self-management will benefit both consumers and service providers immensely. The Internet model also influences the ability of the telecoms workforce to develop applications. As tools and service are developed, that use Internet technologies and standards, telecommunications service providers will have at their disposal a larger number of individuals who can work with web style tools rather than the IN services creation environments of the 1990s.

This section explores the technologies and techniques that have lead to the next-generation network services that will emerge over the coming years. It also gives an overview of the technologies that will allow the telecommunications service providers to create new services.

In telecommunications, voice has always been the predominant application (and some might argue revenue generator, although times are changing in this respect). In the preface, it was discussed that voice has been around in telecommunications for over 100 years as a wire line service. Mobile networks in the 1980s released the tether on voice services. The Internet has and will continue to bring an application revolution to voice services. This first part of the book starts with a description of the way circuit switching functions, and explores the evolution of the circuit switch. It is the evolution of circuit switching combined with the rise in packet data networks that has enabled the so-called convergence of voice and data. The first part of the book continues with an overview of packet networks, and describes the salient information relating to how they have evolved, together with their use for transporting information in the next-generation networks. The first part of the book concludes with how information will be represented and stored in the form of XML and network directories. The combination of all of this technology will lead to the application of voice services as components of more useful services implemented as software executing on open platforms. This service evolution will be explored in the second part of the book.

Each chapter in this part of the book could be expanded to cover a whole book each (and has been by various authors). The brevity of the coverage is an indication of the size of the topics and is meant as an explanation of the key points. I hope the reader will follow up the references quoted for further more detailed analysis of each of the topic areas. This section is designed to highlight the technologies that will and have been influential in delivering the new world of telecommunications.

1

Circuit Switched Technologies

1.1 THE EVOLUTION OF CIRCUIT SWITCHING

The current circuit switched network concept has remained essentially unchanged from the original electromechanical Strowger exchange (see the Preface for an explanation of how this exchange came by its name). At its most basic level the telephone network comprises transmission paths and switching nodes.

The design of a circuit switch is based on the ability to physically create a path (circuit) from one network element to another and to hold this path open for the duration of the interaction (call). The second role of circuit switching is routing, i.e. determining the path to take from the ingress point to the egress point in the network. This can be performed in multiple stages, each switching stage being linked by transmission paths. In the Strowger exchange routing was performed on a step-by-step basis, using the pulsed make and break signalling from the telephone dial to step electromechanical selectors.

Figure 1.1 depicts a simple scenario of an own-exchange call (i.e. a call that only involves one routing and switching stage). This is similar to what would have occurred in the second exchange opened in the UK. Called the *'official switch'*, it was used as a private branch exchange by post office officials at the post office HQ. It can be clearly seen from this simple example how routing is taking place in a step-wise fashion and a physical path is being created at the same time. Clearly, a more complex routing mechanism is required for national and international calls, this is performed by multiple stages of switching and routing connected via transmission paths. In this way the hierarchical nature of the routing

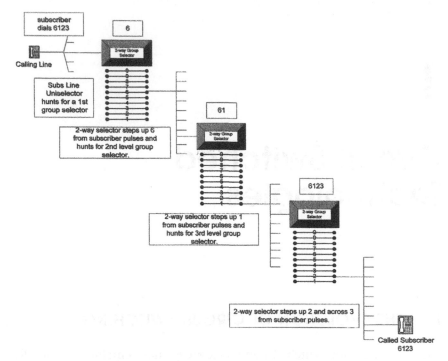

Figure 1.1 Strowger routing scheme –10,000-line, four-digit numbering

and thus numbering plan, local, transit and international evolved. It is on this basis the worldwide numbering scheme evolved!

This example also demonstrates the physical dimensions a Strowger exchange occupied, each electromechanical selector was housed with a number of others in a metal rack. Each of these racks was placed in exchange buildings, in equipment halls. It is safe to say that nearly all Strowger exchanges have now been replaced by electronic exchanges,[1] their replacements being significantly smaller, with greatly increased functionality.

BT crossbar switches (TXK) replaced a number of the Strowger exchanges in the UK. This was a major change, and out went the uniselectors, two-way selectors and progressive control (each switching stage having its own control equipment), to be replaced by a common control and a cross point switch block. Whilst this common control function could only handle one call at a time, its operations were faster than the Strowger staged approach and so a seemingly simultaneous operation could be achieved. A similar evolution occurred in other parts of the world as switch manufacturers released newer switches.

---

[1] The last Strowger exchange was removed in the UK in 1995, if any do remain, they are in developing countries.

In the UK, electronic switching finally usurped the crossbar design in the 1960s with the TXE2 exchange, which used discrete semiconductors in the common control equipment and reed relays in the switch matrix. TXE4 and 4A came along in the 1970s. TXE4A used large-scale integrated circuits in the common control block. This was still in essence a mechanical exchange, with a metallic path from end to end (the TXE4s in the UK network finally disappeared in 1998). It was not until the early 1980s that the replacement of these exchanges with full digital (TXD) equipment, with high-speed semiconductor switch matrices and Stored Program Controllers (SPCs) running software (System X, DMS, AXE10, etc.),[2] finally replaced their mechanical cousins.[3]

It was the SPC and semiconductor switch matrix, which brought about the digitisation of the telephone network. The SPC software could not only perform basic routing capability (which is what it initially performed), but also interpret more complex services. It is instructive to note that this evolution (from the end of Strowger to digital exchanges) occurred over a relatively short period (30 years).

The common elements of a digital circuit switch are shown in Figure 1.2. The elements are SPC, switch matrix, trunk peripherals and Tones & Recorded Announcements (T&RA).

The SPC is the brains of the switch where all the programs that control the call state (finite state machine) reside, along with the signalling, routing, maintenance, charging, and switch matrix control programs.

The switch matrix comes in a number of forms (each switch manufacturer choosing their favourite variation), all of them combine time (also called channel switching) and space switching. Time switching describes how timeslots from an incoming time division stream (see Chapter 2 for a description of timeslots and time division multiplexing) are disassembled from the incoming stream and reassembled on the outgoing stream. This is how '*switching*' takes place. (I will explore switching in a little more detail in a moment, as this is quite a tricky topic!)

The role of trunk peripherals is to terminate the incoming and outgoing time division multiplexed streams. Their role is also to ensure that the streams do not get out of synchronisation, as this would be extremely detrimental (imagine if the timeslots were out of phase, the control software would be connecting the incorrect conversations together!). Timing for the whole of the switch is also derived from information gained from the trunk peripherals. Another component is the Tones and Recorded Announcments (T&RA) source. This component is responsible for

---

[2] All trademarks acknowledged.
[3] The terminology TXE stands for telephone exchange electronic and along with TXS (telephone exchange Strowger), TXK (telephone exchange crossbar) and finally TXD (telephone exchange digital) formed the generic naming of telephone exchange equipment used in the BT network.

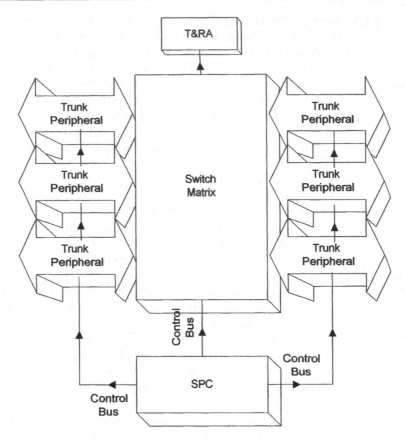

Figure 1.2 Elements of a digital switch

generating call progress tones and announcements that are used to communicate to the caller the status and progress of their call.

Digital switching is performed with two functions: a time switch (see above) and a space switch also known as a timeslot interchanger. In order to understand switching a basic knowledge of the transmission framing in Time Division Multiplex (TDM) is necessary. If you are not familiar with this, then I suggest you turn to Chapter 2 on the transmission infrastructure.

To explain time switching, consider Figure 1.3. A bi-directional path is desired between timeslot 3 on the inbound port and timeslot 27 on the outbound port. We have already established the fact that the trunk peripherals look after synchronisation, so if a switch has all its systems synchronised, then all time division multiplexed streams of voice will be aligned. A time delay must be introduced between the two time division multiplexed streams to allow different parts of each stream to overlap (see Figure 1.3).

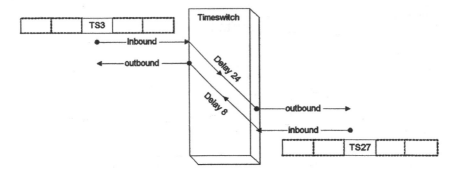

Figure 1.3   Time switch operation

Looking at the figure from left to right, in order for timeslot 3 of the incoming stream to line up with timeslot 27 of the outgoing stream, a delay of 24 timeslots is introduced. From right to left, since the 32-timeslot system repeats frames every 32 timeslots (on a 2.048 Mpbs stream, see Chapter 3 on transmission infrastructure), then a delay of eight timeslots from timeslot 27 is timeslot 3 in the next frame. This process is normally accomplished by the use of Random Access Memory (RAM) to store the bit pattern from each frame and a counter to index the location in the memory.

Two digitally encoded voice conversations can be connected to each other in this way, how do we achieve the connection of hundreds of thousands of connections in an any-to-any way? This is achieved by the use of a space switching stage.

Space switching, is as the name suggests, the act of physical displacement of timeslots (Figure 1.4). Consider a number of time switches aligned on either side of a component that contains a number of crossover points. In order for a timeslot from one time switch to connect to another timeslot in another time switch, a cross point in the component (the space switch) would need to be active at just the right time. The space switch is a timeshared matrix allowing access to all terminations.

A speech sample arriving in a timeslot on the ingress stream is held in a receive store. When the time interval allocated to the cross point being active occurs, the speech sample is read out of the store. The sample traverses the space switch and is written into a transmission (TX) store. When the time for the speech sample to be passed on to the egress stream arrives, it is read from the transmit store.

The final configuration results in a time–space–time architecture (a space–time–space architecture is also possible). At each space switch time allocation slot, the data are read from the input time switch store and transferred across a physical path to an outbound time switch store. This outbound time switch then reads out the data in the appropriate

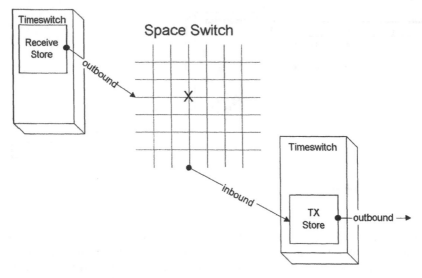

Figure 1.4 Space switch operation

(delayed) outbound timeslot. As you can see this introduces delay at each switching stage.

Thankfully, not very much delay is introduced. A single frame on a 2048 kbps 32-timeslot bearer takes only 125 μs to transmit. Thus, the worst-case delay of a whole frame is only 125 μs. However, if this occurs at every switching stage this delay can soon add up on international links.

Switching is just one component of the connection of telephone calls across a circuit switched network. Whilst digital switching was a very important step in enabling the digitisation of the voice network that has been the enabler for the move to voice and data convergence, the one invariant throughout the history of circuit switching has been routing. Routing is the process of interpreting the digits dialled by one customer into the physical endpoint in the network of the customer they wish to reach and is performed by software running in the SPC of modern circuit switches. Routing is based on a hierarchical routing scheme embodied in the numbering plan. A numbering plan describes the structure for the organisation of the digits customers/subscribers dial to reach other subscribers.

Most people are familiar with the hierarchical routing scheme embodied in the international numbering plan referred to as E.164. The International Telecommunications Union telecommunications (ITU-T) standard specifies a maximum of 15 digits and a geographic hierarchy for the international public telecommunications numbering plan. This numbering plan consists of an international country prefix, followed by a regional number prefix and finally a local number. This hierarchy allows for shortcuts to take place. To call a neighbour, you only need dial the local number without any prefix digits.

The telephone network is divided into local exchanges (incorporating concentration stages that concentrate access network traffic on to links to the local exchange, aka class 5 in the US), transit (or trunk or tandem aka class 4 in the US) exchanges and international exchanges, reflecting this hierarchy of routing. This basic infrastructure remained relatively unchanged all the way up to the 1980s. When the desire to increase the number of services that, the network could offer, whilst reducing the need for increasingly complex software on the SPC was achieved. This was realised by the introduction of the intelligent network architecture (see Chapter 3).

So routing is the process of interpreting the digits from this call plan into a meaningful path through the circuit switched network. Routing is a distributed stage-by-stage process in telephony, with switches at different levels in the hierarchy taking responsibility for different stages in the routing.

By way of an example, consider the number 44-1189-428025. This number has been artificially partitioned into international country code (44), followed by national prefix (1189) and finally the local digits (428025). If a subscriber chose to dial from another country (other than the UK, 44 being the UK country code) then the whole number would be required in order for the telephone network to route the call. If a caller based in Reading (UK) wanted to reach the customer whose number was 428025, then they need only dial this shorter digit string. This is because in the latter case the local exchange that both customers/subscribers are connected to contains sufficient information in the program in the SPC to determine the equipment (and thus subscriber's line) that the number relates to.

If the caller was outside the area they would call 01189 428025 (in the UK). The local exchange that the caller is connected to would have to pass the number up to a transit exchange. The transit exchange could then determine if it needed to pass the call on to another transit exchange, or if it had the local exchange that the number related to directly connected to it. The transit would then pass on the digits to the next exchange in the hierarchy. In order for exchanges to communicate in this way a mechanism for passing the information between exchanges and signalling responses back about the results is needed. This is the topic of the next section.

One final note, the hierarchical approach to routing has been driven by cost as much as numbering plans. The cost of trunking large volumes of copper wire and hence subscriber lines over long distances is significant. The twisted copper pair in most homes is aggregated by local exchange switching centres and carried over multiplexed co-axial and fibre links to the tandem exchanges. We will discover (in Chapter 5) that packet-based voice networks allow us to flatten this infrastructure in a cost-effective way.

1.2 SIGNALLING – COMMUNICATING BETWEEN SWITCHING POINTS

Signalling is the term used to describe the messages that are interchanged between the switching points in order to facilitate the communication of what is known as call progress information. What this statement means is that a mechanism must be in place that allows the communication between telephone exchanges (which are computers in the case of modern digital exchanges) of the dialled digits that a customer dials to reach another customer and a means for errors to be communicated back to the instigating switch (or even customer).

In keeping with the evolution of switching components, the signalling and transmission components have also followed an evolutionary path both at the network edge and in the core of the network.

The edge of the network has slowly undergone the replacement of the loop-signalling interface to a dual tone signalling method (DTMF also known as MF4). Loop signalling is, as the name suggests, a means of signalling the digits dialled by making and breaking a loop circuit between the telephone handset and the local exchange, the loop being formed using the copper twisted pair cable connecting the telephone handset with the telephone exchange. Dual tone multifrequency (DTMF) or multifrequency signalling number 4, as it is also known, is a mechanism that utilises a collection of audible tones arranged in pairs associated with each button on the key pad of a modern telephone handset.

The introduction of digital access signalling at the edge of the network has occurred in the form of a number of different protocols namely:

- Digital Access Signalling System (DASS 1 and 2), a UK centric protocol designed by BT and now largely superseded by DSS1.
- Digital Private Network Signalling System (DPNSS).
- Q.931/I.451 (more accurately known as DSS1 the other two are the call control protocol standards), used for integrated services digital network (ISDN) call set-up signalling for basic and primary rate connections between customer premise equipment and local exchanges. This is also largely being replaced in Europe by Euro ISDN a standard developed by Europe Telecommunications Standards Institute (ETSI).
- Q.SIG, an amalgamation of Q.931 and DPNSS capabilities for signalling for basic and primary rate connections between customer premise equipment and local exchanges and the construction of private networks.

- The US has Telcordia specified ISDN 1 and 2 protocols and Japan has INS-Net defined by NTT.
- V5, a protocol designed for the connection of concentrator switches to local exchanges. It has two versions (V5.1 and V5.2), the second version having more features.

These protocols have allowed the introduction of more sophisticated devices at the edge of the network and through this the evolution of more complex services including circuit switched data services (see Chapter 6). We will not cover these protocols in any more detail, suffice to say they all provide a similar service. That of connecting digital/electronic equipment such as Private Branch Exchanges (PBXs) and Automatic Call Distributors (ACDs) to the public switched telephone network and other private networks. The key point about the move from analogue signalling at the edge of the network to digital signalling is the increase in services and facilities that can be supported, and for example the ability for end devices to communicate with each other, using the public switched telephone network as a packet data network for carrying those messages.

One facility that makes good use of this is route optimisation. When two private exchanges (PBXs) are connected together through a number of other exchanges (as transiting exchanges). One of the parties in the call wants to redirect their end of the call to a third person and hang up (transfer the call). The route the new call takes can be optimised by dropping the path of the call back through a number of the intermediate exchanges until it passes through the minimum number of exchange links. This facility is provided by signalling messages that pass between the edge PBXs and intermediate nodes to establish the new route.

The core network signalling, in concert with the access network signalling, has evolved from analogue-based signalling in the form of:

- Loop disconnect (see above) this is a form of direct current signalling that is only effective over circuits up to about 2 km.
- E&M, stands for ear and mouth signalling, this is a two-way signalling mechanism, ear being the receive signalling and mouth the transmit signalling.
- DC2 and DC3, use current pulses to signal digits and trunk seizures between exchanges.
- AC8, AC9, AC11 and AC12, these are all what are referred to as out-of-band and in-band signalling systems. They use frequencies of sound outside those normally permitted for voice (artificially filtered) and sounds inside the voice range.
- MF2, like its cousin MF4 (see above), was used for speeding up the transmission of decadic digits between trunk exchanges by encoding the digits as a set of in-band tones.
- R1 and R2. Signalling systems R1 (North America) and R2 (Europe) are used for inter-register signalling. Inter-register signalling

(between trunk exchanges) uses MF in-band pulse signalling at frequencies of 700–1700 Hz, in 200 Hz steps, for the transmission of address information. Line signalling is performed in TDM systems as a set of bits (normally in channel 16 of a 32-channel system and using bit robbing in US 24-channel systems – see Section 2.2).

To a digital packet-based signalling system called signalling system number 7 (SS#7). I can hear what you are thinking, "What happened to the other SS#x?" SS#4, 5, 6 are international analogue signalling systems specified by the then CCITT (ITU) in the early 1960s. I will not cover any of the analogue signalling systems in this text because, whilst their importance is recognised, they are largely being/have been replaced by packet-based digital signalling in the form of SS#7 in the Public Switched Telephone Network (PSTN).[4] The move over to packet-based systems is because of a number of reasons. In the previous (analogue) signalling systems mentioned:

- a direct relationship exists between the telephony traffic route and the signalling (in packet-based signalling the messages can follow independent paths to the telephony traffic);
- only telephony data could be signalled (in packet-based signalling, network management messages, statistics information and fault reports are all carried over the signalling system);
- the number of messages are limited;
- the signalling transfer of messages is slow; and
- equipment is inefficiently used because it was generally dedicated to a specific route.

This brings us neatly on to the now universally accepted packet-based signalling system, SS#7.

Overview of Signalling System Number 7 (SS#7)

This section covers the signalling protocols set out in the ITU-T[5] standardisation sector specifications known as the Q.700 through to Q.775 series of recommendations. The term recommendation is an interesting one in that it implies they are not compulsory, however, without almost universal adoption by telecoms equipment manufacturers and network operators nothing in the telephone network would work, and would not have moved beyond operator-connected calls.

---

[4] I'm certain to get remarks over this point, as I'm sure a significant amount of international analogue signalling still exists! The important point is the move to digital signalling and what that enables.
[5] Formally known as the Consultative Committee International, for Telephony & Telegraphy (CCITT), the conversion took place on 1 March 1993.

The ITU-T recommendations have peer specifications from the American National Standards Institute (ANSI) and Bellcore (now Telcordia Technologies) for the North American variants of SS#7, also regional variants exist, for example Japan, and in the UK, the Telephony User Part (TUP) protocol is known as IUP and formally as BTUP. A good reference for a more detailed study of these specifications and SS#7 in general can be found in [RUSS]. By completing this section, you will be rewarded with a good foundation to take you through to how and why intelligent networks work.

In order to discuss SS#7, an understanding of the International Standards Organisations – Open Systems Interconnection (ISO-OSI) seven-layer model is not necessary (but helpful). The specifications for SS#7 were put together before the ISO model and thus the protocol stack only has four levels (not seven like the OSI), message transfer part (levels or layers 1–3) and user parts layer 4 (the Q.700 document describes this structure). These roughly equate to the OSI layers 1 through 5 (physical, data link, network, transport, session). The presentation and application layers of the ISO stack strictly speaking are not present in the SS#7 stack, however, some of the presentation layer functions are included in the SS#7 transaction capabilities layer. The ISO-OSI transport layer is not applicable to Message Transfer Part (MTP) layer 3, as MTP does not make use of connection oriented services. However, some applications make use of a connection oriented service present in Signalling Connect Control Part (SCCP), hence why in Figure 1.5 the ISO transport layer overlaps the

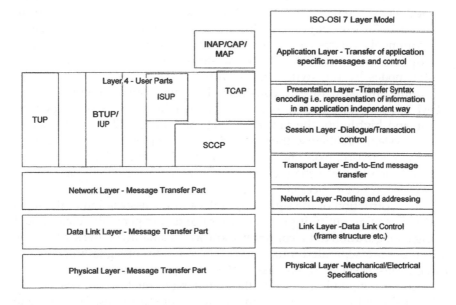

Figure 1.5 SS#7 four-layer protocol stack

bottom portion of the SS#7 SCCP layer. Figure 1.5 shows the SS#7 protocol stack in comparison to the ISO-OSI seven-layer model, this is only to give a perspective on how the SS#7 protocols function and doesn't represent a direct relationship between the two.

Figure 1.5 shows INAP, CAP and MAP protocols in layer 7 (application layer). These are included in the figure for completeness and are discussed in Chapters 3 and 4, respectively. INAP stands for intelligent network application protocol and provides the functionality to support enhanced services in fixed networks. CAP stands for Customised Applications for Mobile Networks Enhanced Logic (CAMEL) Application Part and MAP for Mobile Application Part, both protocols for mobile networks that provide enhanced application functionality specifically with mobility in mind.

Message Transfer Part

The Message Transfer Part (MTP) covers layers 1 through 3, we shall take these in turn from layer 1 through to layer 3.

Layer 1 covers the physical presentation of the signalling, this specifies either a V.35 interface or a single TDM slot (DS0A – bit stolen signalling channel see Chapter 2 on framing for an explanation of bit robbing, or DS0C – clear channel 64 kbps). In most networks (fixed and mobile), the signalling is carried over a TDM slot, as these are readily available from the transmission infrastructure (see Chapter 2).

It is common in the telephone network for the signalling channels to be carried alongside the voice channels as a clear channel 64 kbps bearer in timeslot 16 (in Europe) of a 32-timeslot system. The individual timeslots are commonly referred to as bearers and/or channels. The signalling is multiplexed into timeslot 16 from the signalling software on each of the switch nodes by the switch matrix function. A complete packet-based signalling infrastructure is constructed in this way, embedded in the TDM transmission bearers between the switches. This is an important point and highlights the fact that MTP, and in fact SS#7, is a packet-based system. In theory, there could be a completely separate packet switched network, from the circuit switch voice connections. The use of the multiplexed voice channel systems to carry the signalling packets is a convenience, not a necessity.

The construction of the signalling links and the association of the links to the signalling entity it provides a service to, are generally constructed using configuration data (commonly known as datafill) on the switch. The link types are separated into modes. These modes of signalling come in three forms: associated, non-associated and quasi-associated.

Associated signalling is when the signalling in the timeslot has a direct relationship with the speech channels on the link it shares. This mode of signalling is commonly used to signal messages about an analogue term-

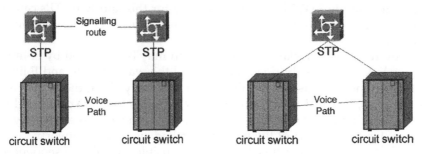

Figure 1.6 Signalling modes

inal on a multiplexor. In the case of a 32-timeslot system, timeslot 16, that contains 8 bits, is subdivided into two 4-bit parts. The lower 4 bits are associated with information about the voice connections in the timeslots below 16 and the upper 4 bits contain signalling information for the upper 15 speech channels.

Non-associated signalling uses a separate path to carry the signalling information from the voice bearers that are used to carry the voice channels (Figure 1.6). In this instance, a separate packet switched network connection is essentially constructed. It is at this point we will introduce the term Signalling Transfer Point (STP). A STP is the SS#7 equivalent of a router. In most cases, the STP is an adjunct function embedded in the circuit switch, composed of hardware and software components (in the SPC). The circuit switch is used to connect the timeslot 16s through to a number of ports on the switch that have special trunk peripherals that connect the signalling channels to the control bus, and thus the signalling messages to the software on the SPC (see Figure 1.2). In non-associated signalling, messages can actually transit through a number (one or more) of intermediate routing points before reaching their destination.

In quasi-associated signalling, the two signalling endpoints are connected to the same STP and the signalling path is separate to the voice path.

In order to provide capacity and redundancy for signalling, a number of signalling links (up to 16) can be grouped together into what is called a *linkset*. If more than one link is provisioned into a linkset, then the messages are load shared across the links. These linksets are then grouped into routes, if all the linksets the group can be used to reach a particular signalling destination. Routesets are then defined as collections of routes. This grouping is to allow the STPs to find alternative routes to a destination for example during a failure of equipment.

Clearly, a connection to only one STP would be unfavourable from a resilience perspective and with reliability as one of the key design criteria it would be unwise to only be able to connect the signalling point in a switch to a single STP. The standards allow for this and two STPs can be combined into a pair, with associated paired signalling linksets from a switch node.

If you read the associated standards, you may be confused by A-links (access links), B-links (bridge links), etc. all the way up to E- and F-links (extended and fully associated links). Do not worry about these, they are all the same, it is just which equipment endpoints they are connected to that makes them different, they function in the same way as an overlay of bearers for the support of the packet switched signalling network.

MTP layer 2, the data link layer is responsible for reliable sequenced delivery of all SS#7 messages, and it achieves this on a node-to-node basis. The key feature of MTP is reliability. The specification of MTP goes to extreme lengths to provide a reliable transport mechanism for user part protocols. MTP layer 2 provides both sequence numbers for the messages it sends and a cyclical redundancy checksum for the packet. Any messages found to be in error or lost are retransmitted by MTP layer 2.

Layer 3 of MTP (the network layer), is responsible for the routing of messages, message discrimination and message distribution. Message discrimination determines if the message is for a local layer 4 protocol or if it is for another node. If it is a local message, then it passes the message to the distribution function for delivery to the appropriate layer 4 user part (protocol, for example ISDN user part (ISUP)). If the message is destined for another node, it passes the message to the routing function. Layer 3 also performs the vital role of supporting network management. Network management in MTP constantly monitors the links for errors and congestion. MTP has special messages called signalling units and in particular a message signalling unit. The message signalling unit message is used to automatically reroute messages around failed links by instigating what is called a changeover. Changeover is in essence a way for the signalling transfer points to inform each other of a failed signalling and start sending signalling information down the partner link of a signalling linkset.

Addressing in layer 3 is determined by what are called point codes. Every signalling function in every switch in all the inter-connected SS#7 networks in the world has a point code. In this way, MTP layer 3 is comparable to the Internet Protocol (IP) and a point code comparable to an IP address.

User Part Layer

Layer 4 is where the call control signalling and some application signalling are located. For example ISUP is a call control protocol which specifies how connections are set up and torn down. ISUP was designed to work in concert with the Q.931 digital access signalling system and allows for the connection of both telephony and switch data services amongst other services such as call forward and calling line identification (called Automatic Number Identification (ANI) in the US).

TCAP (Transaction Capabilities Application Part) is an application protocol for accessing databases. You see TCAP being used in intelligent networks to carry INAP messages (see Chapter 3 for more on intelligent networks). TCAP is a complex protocol and implements a lot of the services present in the ISO session layer, and some of the ISO application layer functions.

TCAP makes use of the services of the protocol known as SCCP (Signalling Connection Control Part). This protocol provides another layer of addressing beyond point codes, *subsystem number*. This subsystem number is used to reference a particular instance of a service or a specific database. SCCP also provides services to the TCAP layer more akin to services offered by the ISO transport layer, namely: a connectionless service (sequenced and un-sequenced), a connection oriented service and a flow control connection oriented service. SCCP provides the end-to-end service for messages that MTP does not, as such SCCP supports message transfer and routing between non-telephony applications, i.e. database lookups.

There is a special routing stage that can take place in the SCCP layer that is very powerful. This routing capability is called *Global Title Translation* (GTT). What GTT does is to associate a service request code (an 800 number for example) with a point code and subsystem number. This routing capability takes place in the STPs. Why is this important? Well GTT means that fault tolerance and load sharing across service points (or databases) can take place without the invoking switch being aware of it happening, you will see the importance of this in Chapter 3, on intelligent networks.

We are going to take a brief interlude from switching and signalling to cover transmission systems, and then we will pick up advance telephony services in the form of intelligent networks in Chapter 3 and explore the use of SS#7 for more advanced services.

2

The Transmission Infrastructure

2.1 INTRODUCTION

We pass by transmission on the way to the Intelligent Network (IN) because it is important to understand how all the switching and signalling nodes (not to mention voice links) are connected together. Like the last chapter, we will pass briefly through analogue transmission first as a means of exposing the desires for digital transmission. Then move on to digitisation of speech and look at how large volumes of calls are economically carried across the world.

Analogue voice signals can suffer from a number of interference problems ranging from degradation due to distance through to external signalling inducing noise into transmission. One of the big culprits for 'noise' on the line is cross talk. Cross talk occurs when a number of transmission systems are carried through the cabling trunk in close proximity. External noise in the form of interference from electricity mains cabling and other electronic equipment conspire to reduce the speech to an unintelligible hiss. It is this reason as well as economies of scale in amalgamating voice connections together into large transmission systems that has brought about the desire to digitise the analogue signals produced by the human voice. Digital signals are less prone to interference as interference only indirectly affects the signal. The original analogue signal must be decoded from the binary representation, as long as the accuracy of the binary representation of the signal is maintained the original signal can be regenerated immune from noise.

2.2   VOICE DIGITISATION

Speech using the handset concept created by Alexander Graham Bell is an analogue signal. It is a continuous signal, varying in amplitude and frequency in sympathy with the compression waves created from the human voice box in the process of producing speech. In order to represent this accurately as a digital signal, it must undergo three processes: filtering, sampling, quantising and encoding (these latter two being a single step).

Filtering is the process of eliminating frequencies beyond a certain range. The audible range is from around 20 Hz to 20 kHz. The majority of information from a human voice is actually present between 300 and 4000 Hz (the human ear's maximum range of sensitivity is from around 1 kHz up to around 5 kHz). This frequency range is used in most telephone systems around the world to convey the voice of the person speaking. The reason for this filtering probably goes back to analogue transmission of multiple voice signalling using a technique known as Frequency Division Multiplexing (FDM). Basically FDM uses collection of frequencies and modulates them with the original voice signal. The modulation changes the base frequency up and down in sympathy with the changes in the voice signal. Clearly the broader the spectrum of frequencies present in the voice signal the more the base frequency would vary, this would have the effect of limiting the number of separate voice 'channels' that could be carried on an FDM trunk. By artificially constraining the voice to a limited range (300–3400 Hz in Europe and 200–3200 Hz in the US) of frequencies the capacity of the FDM trunk could be increased.

Sampling is the process of taking discrete moments in time and measuring the value of the amplitude (loudness) of the audio signal. The sampling rate used for telephony is 8000 times per second. Why this rate, a mathematician named Nyquist proved a sampling theorem that states a signal's amplitude must be sampled at a minimum of twice the highest frequency of the signal. Since the maximum frequency allowed by the filter is 4000 Hz, then a minimum sampling rate of 8000 times per second is required for a 4000 Hz signal.

Finally, to convert the sampled signal called a Pulse Amplitude Modulated (PAM) (Figure 2.1) signal into a set of binary pulses, the PAM sample must be given a discrete value. This value, in the scheme used for the circuit switched telephone network is an 8-bit binary value in the range of +127 to −128, thus including zero a range of 256 distinct amplitude levels.

Ideally, more levels (in fact an infinite number) are necessary to truly represent the signal. Eight bits are seen as a sufficient compromise for voice signals in telephony. For example in compact disc (CD) recordings the signal is sampled 44,100 times per second, and a range of values of 65,536 are used to represent the sound. To think about what is happening

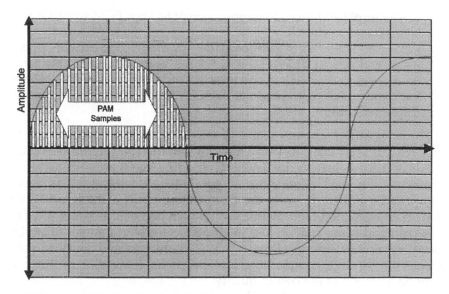

Figure 2.1   PAM sampling of an analogue signal

here, as the sample rate is increased and the number of values (precision) used to represent the analogue waveform increase, the digital sampling is tending closer to the actual analogue signal it is trying to represent. At infinity the digital pulses become a continuous waveform, essentially the original analogue waveform!

This results in what is called a Pulse Code Modulated (PCM) signal. The encoding (actually called companding, because this signal is first compressed from 12 bits to 8 bits then expanded in the decoder to 13 bits) part is actually performed using an approximate logarithmic scale (the step value is doubled for every doubling of the input level); this is to avoid unacceptable errors (due to approximation) for small amplitude signals and incidentally is how the human ear resolves sounds. The scale used is either an A-law (European) or a μ-law (North America) companding. If you would like more detail on this topic, then I refer you to [HALS, HERS], both of which cover this topic well.

A-law and μ-law are actually an International Telecommunications Union telecommunications (ITU-T) standard known as G.711 encoding. Other encoding techniques exist for lower bit rate transmission such as:

- ITU-T G.726 – adaptive differential PCM (ADPCM), with bit rates as low as 16 kbps. ADPCM reduces the bit rate by dynamically changing the coding scale and only encoding differences from one sample to the next.

- ITU-T G.728, 16 kbps low-delay code excited linear prediction (LD-CELP). This algorithm uses a look-up table to extract values for the

voice sample and provide compression of 64 kbps PCM signals down
to as low as 16 kbps.

- G.729 8 kbps conjugate-structure algebraic-code-excited linear
prediction (CS-ACELP). ITU-T G.729 has been added to in the form
of annexes to the original specification and vendors have implemented these, resulting in products such G.729a, referring to an annex A
implementation.

These encoding schemes won't be examined further in the text here,
and readers who want to know more are referred to the relevant standards or [HERS]. Just one final note on encoding that is important when it
comes to the next generation of packet-based telephony. As more
compression is introduced in the companding scheme, the resulting bit
stream encodes more information and as such becomes more sensitive to
bit errors and frame/packet loss. As the telecoms world moves to implement Voice over Internet Protocol (VoIP) solutions and looks to gain
economies through compressing the speech for transmission, this can
only be achieved at the cost of quality.

The bit stream resulting from the companding process can be combined
with many others to form a single serial bit stream (Figure 2.2). The
combination of each discrete stream is created by time slicing each one
in turn on to the serial transmission line. Since we are producing 8 bits of
data, 8000 times per second, the resulting bit rate is 64 kbps. Each 8-bit
sample occurring ever 125 μs. This means in order to create a system
containing 32 timeslots, each individual stream must be sampled every

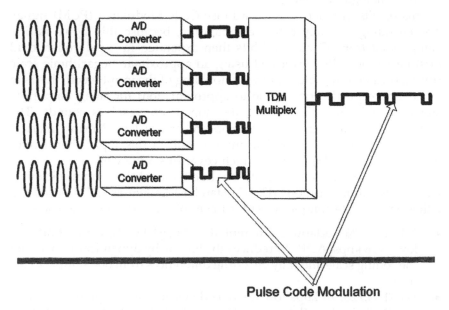

Pulse Code Modulation

Figure 2.2   Construction of a digital time division multiplex bit stream

125 μs, resulting in a gross bit rate of $8000 \times 32 \times 8 = 2.048$ Mbps. This stream is known as a primary rate stream or an ITU-T E1 system.

The final stage in the process of producing a time division multiplex stream that can be transported over a reasonable distance on copper co-axial cables is line encoding. Line encoding serves the following purposes: to create few low frequency components, create no zero frequency components, encode timing information and finally to provide a means of monitoring for errors caused by loss or noise on the transmission line.

The encoding technique explored here is called high density bipolar 3 (HDB3), which is an ITU-T specification used in E1 2.048 Mbps transmission lines (see ITU-T G.703 for more information). This encoding scheme ensures that there are no long streams of zeros or ones present on the transmission line. It achieves this by the following rules: binary ones are transmitted alternately as either a positive voltage or a negative voltage (a mark), a binary zero is transmitted as a zero voltage. This is essentially what is called Alternate Mark Inversion (AMI). A number of other encoding schemes exist to meet the same purpose (Non-Return to Zero (NRZ), AMI, Manchester coding, Zero Code Suppression (ZCS), Bipolar with 8 Zero Substitution (B8ZS) and Zero Byte Timeslot Interchange (ZBTSI), to name a few). B8ZS is widely used in North American transmission systems, whilst HDB3 is used outside North America.

For HDB3 encoding, in any sequence of four consecutive zeros, the final zero is substituted on the transmission line with a mark of the same polarity (+ve or −ve) as the previous mark (this is called a bipolar violation). If a long stream of zeros were present then clearly every fourth zero would be replaced by a mark on the transmission line, however, successive violation marks of this nature are of opposite polarity. Why, since applying the rule above is each bipolar violation of the same polarity as the previous mark? This is because where successive violations would be of the same polarity, a *balance* mark is inserted by setting the first digit of a sequence of four zeros to be a mark of the opposite polarity of the previous mark. Lost, maybe Figure 2.3 will help.

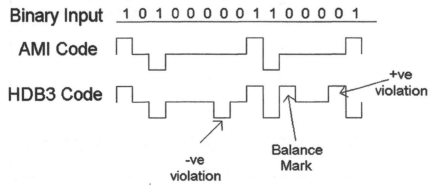

Figure 2.3 HDB3 coding

These violation marks and balance marks also serve the purpose of embedding a timing signature into the transmission. This is achieved through the frequent transitions of the signal.

Timing (synchronisation) was mentioned earlier in the section on switching where it was pointed out that the importance of timing was to ensure conversations between two people could be accurately connected together. In the discussion on HDB3 we saw that timing is coded into the line signal by ensuring adequate pulse density. The encoding also serves as an error detection mechanism; any bit errors created by interference can be detected, and in some instances automatically rectified in the receiver. The final piece of the puzzle for ensuring the channels of the conversation you want to connect are the correct ones is in the use of framing.

Each collection of 32 timeslots of the ITU-T E1 Time Division Multiplex (TDM) (24 in a North American T1) serial pulse train is called a *frame*. This frame is combined into multi-frames. In fact 16 frames in all make up a multi-frame (Figure 2.4).

The 32 timeslots are divided into timing slots, voice bearers and signalling bearers. Timing is embedded in the framing structure by placing special sequences of bits (bit patterns) into timeslot 0 in consecutive odd and even numbered frames and one special multi-frame alignment pattern in timeslot 16 of frame 0. All the other timeslots: 1 through 15, 17 through 31 carry the speech channels and timeslot 16 in all the other frames other than zero carries signalling information. Thus synchronisation of the speech channels is achieved.

In the North American Digital Stream 1 system, a different scheme is

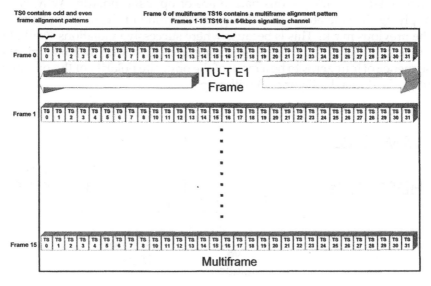

Figure 2.4 E1 framing

used to carry 24 voice channels in a frame. Signalling is carried in this scheme by borrowing a bit out of each of timeslots 6 and 12, respectively (as opposed to a dedicated timeslot), this scheme is commonly known as bit robbing.

To finally finish off the topic of timing in a TDM network, the type of timing is expressed in three words: synchronous, asynchronous, plesiochronous. Clearly, we can discount asynchronous, since the whole topic so far has concentrated on synchronising voice channels. Synchronous can be discarded since this implies all the clocks in all the exchanges are in complete synchronism, which is not the case, because a single clock source connected directly to all exchanges would have to be present. Plesiochronous on the other hand means '*nearly*' synchronous and that is just what a TDM voice network is. In fact a hierarchy of timing is present in a TDM voice network. With the master source generally being a caesium clock, this clock source is rippled down through the international exchanges to the transit/trunk exchanges and finally to the local exchanges, using the timing mechanisms embedded in the bit streams carrying voice and signalling discussed above (see ITU-T G.810, G.811 and G.812, if you are interested). All this is in place to guarantee that once a connection is set up between one endpoint and another, it remains connected with little or no change in delay (variance in delay is commonly referred to as *jitter*) and no loss of clarity, through the loss of voice samples. We will see later when we discuss packet-based voice communications just what a legacy this is.

Therefore, that is how a collection of 30 speech channels is placed on a serial transmission line. What we do not want to do is put lots of single E1 or T1 links in the ground. This clearly would not be economical. Therefore, we need a means by which we can combine the E1s into a collection of E1s for economic transmission. That is where PDH, SDH, ATM, DTM and DWDM come in (each of these acronyms will be explained in turn in the following sections).

2.3   PLESIOCHRONOUS DIGITAL HIERARCHY

In the previous chapter, we covered how individual digitally encoded voice samples are combined into 32 timeslots 30 channels of voice E1, or 24 channels of voice T1. To economically carry lots of these voice bearers across the world, and maintain the timing of the switched network Plesiochronous Digital Hierarchy (PDH) was created. Across the globe three systems of PDH exist, one in North America (where it was first invented by Bell Labs, now part of Lucent), one in Europe and one in Japan. The three variants all share one thing in common, they all specify five levels of multiplexing.

Simply put the PDH network is an aggregation of E1/T1 bearers in higher bit rate systems by multiplexing the lower order E1s/T1s together. The ITU-T specifies the following higher order bearers: E2 (8.448 Mbps), E3 (34.368 Mbps), E4 (139.264 Mbps) and E5 (565.148 Mbps).

The North American system refers to each level as digital streams (DS). DS0 is the bottom and represents a single 64 kbps channel, DS1 is the 24-channel system (also referred to elsewhere in the book as T1. T1 is the colloquial term used that was actually a reference to the four-wire transmission system). DS1c is two DS1s, and DS2 two DS1cs, DS3 (also colloquially referred to as T3, although strictly speaking there is no such thing as a T3) is seven DS2s.

The smart reader will spot the bit rates in the European PDH are not exact multiples of 2.048 Mbps, why is that? In the same way the E1 combines framing information with the voice and signalling, we need to do the same with the higher order bearers. These additional signals are because each of the TDM streams all have a slightly different timing source. This means we need to compensate for this. This is achieved by clocking at a slightly higher bit rate than the sum of the lower order bearers (called tributaries and sometimes colloquially referred to as *tribs* by engineers). Any unused bits are filled with what are called justification bits and alarm signals to indicate failures such as loss of synchronisation. This results in a collection of nearly synchronous bit streams, hence plesiochronous.

This type of network infrastructure still exists today, but is slowly being replaced by our next technology Synchronous Digital Hierarchy (SDH). More on that in a moment, the obvious question is why replace it? The less than obvious answer (maybe) is cost. Consider the collections of E1s all aggregated on to higher and higher rate bit streams, all slightly displaced within the higher order streams by varying amounts of justification bits. This makes it impossible to determine (without de-multiplexing) where each individual E1 starts and ends.

Why is having a hierarchy of multiplexers a problem? Remember when we discussed switching, each switching stage has trunk peripherals, these all terminate E1s or T1s. The switching stage then interchanges timeslots on these bearers to connect speech channels together. The Public Switched Telephone Network (PSTN) is a web of switch nodes connected together by transmission infrastructure. In the case of PDH this is a collection of E5 bearers between major cities (trunk exchanges). Each time a telephone connection is required between cities, the multiplexed collection of E1s will need to be de-multiplexed to get access to the voice channel. In each exchange building there is a large investment in multiplexing equipment. This also extends all the way to the edge of the network in what are called Points of Presence (POP), which the network operator uses to bring customer connections into the network. Figure 2.5 shows the situation.

To alleviate the problem of large multiplexer hierarchies and to increase the rates of multiplexing, SDH was created, to capitalise on the significant

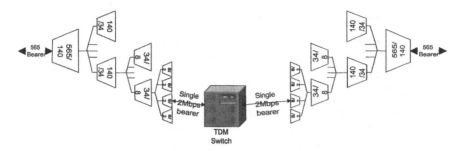

Figure 2.5 PDH infrastructure

installed base of copper twisted pair, co-axial cables and fibre optic cables left as the 30-year heritage of PDH (first-generation transport system).

That was rather a whirlwind tour of PDH, but explains the basics of the infrastructure that grew up with digital switching over the last 30 years. If you would like to know more, then I recommend you consult the ITU-T G.702, G.703, G.704 and G.706 specifications.

2.4 SYNCHRONOUS DIGITAL HIERARCHY AND SYNCHRONOUS OPTICAL NETWORKS

Synchronous Optical NETworks (SONET) was initially specified by American National Standards Institute (ANSI) and Bellcore (Telcordia) in the early 1980s and taken up by the ITU-T to form the SDH standards. It wasn't until the early 1990s that equipment started to appear. This description belies a fraught process; at one point the SONET proposed framing structure of 13 rows by 180-byte columns; whilst the SDH speci-fication advocated a 9 rows by 270-byte column. These differences were brought about by the differences in the basic blocks of transmission that needed to be carried, namely T1 (1.544 Mbps) and E1 (2.048 Mbps). Finally, both standards bodies saw sense and settled on a 9-row frame, wherein SONET became a subset of SDH.

What is the main differentiator between SONET/SDH and PDH? The two words in its name give it away – synchronous and optical. Unlike PDH, synchronisation is maintained throughout the network, and large capacity transmission paths are created over fibre optic cables. Some might argue that SDH is no more synchronous than PDH, whilst this could be argued based on the fact that the bearers/tributaries carried in SDH frames are not in synchrony (stuffing bits are used to cater for fluc-tuations in the timing of the tributaries – this is explained a little later when the SDH frame structure is explained) the rest of the network is.

The subtlety of the difference between SDH and SONET is shown in Table 2.1. SONET describes Synchronous Transport Signals (STS) and

Table 2.1   SONET and SDH transmission rates and names

SONET	SDH	Bit rate (Mbps)
STS-1/OC-1		51.84
STS-3/OC-3	STM-1	155.52
STS-9/OC-9		466.56
STS-12/OC-12	STM-4	622.08
STS-18/OC-18		933.12
STS-24/OC-24		1244.16
STS-36/OC-36		1866.24
STS-48/OC-48	STM-16	2488.32
STS-96/OC-96		4876.64
STS-192/OC-192	STM-64	9953.28

Optical Carrier (OC) signals as the basic building blocks. An STS is nothing more than a framing structure that defines how data can be multiplexed into the SONET hierarchy. SDH defines Synchronous Transport Modules (STM) – again a framing structure. The lowest rate SDH frame starts at 155 Mbps, whereas SONET starts at 51 Mbps. This difference in base rate also means that the base frame and multiplexing structure are also different to accommodate the difference in bit rates.

This results in an STS-1 having a frame size of 90 bytes by 9 rows, and an STM-1 a frame size of 270 bytes by 9 rows. In line with the previous section on PDH, because the speech is sampled at 8000 times per second, each rectangular 9 rows by x-byte columns are transmitted every 125 μs. Resulting in the gross bit rate of $9 \times 270 \times 8 \times 8000 = 155.52$ Mbps and $9 \times 90 \times 8 \times 8000 = 51.84$ Mbps, for STM-1 and STS-1, respectively.

SONET and SDH achieve synchronisation by the use of a single master clock. In order to accommodate plesiochronous tributaries, pointers are used in the frame header. These pointers are values in the frame that indicate the offset of the particular multiplexed unit (E1, T1, E3, etc.) in what is referred to as the *payload*. Each frame consists of a header area and a payload area.

The header contains in SDH an area referred to as the *section overhead*. This section overhead is subdivided into: *regeneration section overhead*, *administration unit pointers*, where the pointers to the payloads reside, and finally the *multiplex section overhead*.

In SONET the header is referred to as a *transport overhead*. This is subdivided into *section overhead* and *line overhead*. This is where I end my discussion on SONET and concentrate on SDH (Figure 2.6 shows the frame structures). For a more detailed coverage of SONET see [BLACK1].

The payload area of the frame in SDH contains the multiplexed streams and a *path overhead*. The path overhead is essentially an area where monitoring information can be stored in the frame, for example to monitor the

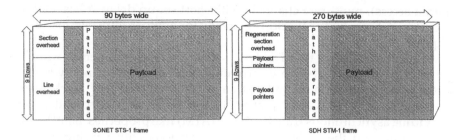

Figure 2.6   SDH and SONET frames

bit error rate of the container (see later). The multiplexed streams in the payload can be assigned in different ways according to the multiplexing taking place.

The lower level multiplex streams are called tributaries, like their PDH cousins, these are placed in a *container* and the containers are combined with any stuffing bits and the path overhead bits to form a *Virtual Container* (VC). SONET refers to these as Virtual Tributaries (VTs), but the concept is the same. The stuffing bits are to allow for variations in rate of the PDH stream within the container. Each STM-1 frame can contain multiple VCs/VTs.

SDH is now the prevalent technology installed as the transport infrastructure in many telecoms networks. A more complete overview of SDH and SONET can be found in [SILL].

2.5   DYNAMIC SYNCHRONOUS TRANSFER MODE (DTM)

The Dynamic Synchronous Transfer Mode (DTM) technology is a hybrid of time division circuit switching techniques and dynamic reallocation of timeslots. It claims to provide the best of both packet switching and circuit switching. DTM is designed as a new technology aimed at the metropolitan area and campus networks. Its originated in the early 1990s, in a Swedish research programme called MultiG, whose aim was to research multimedia applications over gigabit networks. DTM grew out of the work of six people, Lars Gauffin, Erik Hagersten, Christer Bohm, Marcus Hidell, Lars Ramfelt and Per Lindgren [PERS]. This work was later taken up by Net Insight (where indecently Lars Gauffin, Christer Bohm and Per Lindgren now work), Ericsson and Dynarc AB, and products exist, for example, Allied Telesyn produce a number of DTM products, as do Dynarc (obviously).

DTM is a shared medium technology (like the original Ethernet) all stations on the DTM network have access to the fibre optic cable that carries the information. DTM provides multicast (by virtue of the shared media access), multi-rate channels with short set-up delay. It has a lot of promise, but maybe the world isn't ready for another transmission technique?

We established early that TDM has some very useful properties when it comes to carrying real-time information, low latency and non-variable latency and separation of traffic streams creating a security mechanism (see Section 2.2). DTM aims to capitalise on these properties, whilst also providing asynchronous transfer mode (ATM) like properties: bandwidth on demand and the ability to serve applications with varying quality requirements. It also aims to avoid the static bandwidth allocation problem (pre-provisioning of streams) that SONET and SDH rely on, and the waste of capacity created in SDH by having to over provision links based on the multiplexing hierarchy.

Bandwidth in DTM is allocated in discrete chunks of 512 kbps, up to the capacity of the fibre that carries it. It is actually carried in 125 μs cycles of 64-bit timeslots. These slots are further separated into control slots and data slots. The slots themselves cycle around and are given numbers. To create an individual data carrying channel, then a number of slots are allocated in multiples of 512 kbps (essentially one slot at a time 64 bits every 125 μs). So for example in a fibre capable of carrying 622 Mbps, this equates to around 1200, 512 kbps slots, and a 10 Gbps fibre can carry around 19,500, 512 kbps timeslots.

If DTM is used in conjunction with wave division multiplex equipment then potentially huge amounts of bandwidth can be segmented into 512 kbps chunks. I leave it to the readers' imaginations to work out what to do with this capability.

For more on DTM I suggest you visit the DTM forum website at http://www.dtmforum.org.

2.6   WAVE DIVISION MULTIPLEX

Wave Division Multiplex (WDM) is a means of increasing both the lifetime and capacity of existing multimode fibre optic cables. The fibre laid down in the 1980s to carry PDH is, through the techniques of multiplexing multiple wavelengths of light on to a single fibre, being used to carry much greater volumes of telephone calls and data traffic.

What is wave division multiplex? WDM is a technology that initially used two wavelengths of light in the 1300 and 1500 nm wavelength (lambda – λ) windows to transmit a single digital signal, the bandwidth available in these two channels roughly equated to 10 Gbps. This effectively created a two-channel system. A four-channel system called wide WDM was later created and now systems consisting of over 100 optical channels are possible, this is referred to as dense WDM, and means a single fibre can transport terabits of information.

Lasers are used to generate the specific lambdas (wavelengths), and optical multiplexer – effectively a prism is used to multiplex the signals on the fibre. At the far end a de-multiplexer separates out each of the lambdas. Optical signals can be added and removed from a system using Optical Add-Drop Mulitplexers (OADMs).

WDM is compatible with SDH, ATM, DTM (and for that matter Ethernet, work is in progress to standardise the carriage of 10 Gbps Ethernet over WDM). It is not a competing technology, more an addition to the previously mentioned technologies as a means of getting even more capacity out of the existing fibre optic infrastructure.

You may hear of a glut of capacity being present in the world, and that this is driving down cost of wide area links. Whilst this is true, things rarely stay still for long, as we continue to roll out higher and higher capacity at the edge of the network (see Chapter 6 on access technologies) this core glut will soon be utilised. A lot of complexity is present in the management and provisioning of equipment to perform the really high rates of 120 separate channels on a dense WDM system, the really important point about WDM is the capacity increase it creates from the existing infrastructure without having to resort to expensive civil engineering schemes to dig up pathways and sidewalks.

3

Intelligent Networks

3.1 INTRODUCTION

The Intelligent Network (IN) standards are divided into the Telcordia standardisation called Advanced Intelligent Networks (AIN) and the International Telecommunications Union telecommunications (ITU-T) IN standards capability sets. The Telcordia standards are used mainly in North America, whilst the ITU-T standards are relevant in pretty much the rest of the world. Other work has taken place notably by a group known as TINA-C, the Telecommunications Information Networking Architecture Consortium, looking at the longer term architectures for distributed intelligence in telecommunications networks.

The early implementations of IN were based on a database performing number translation of non-geographic numbers such as the US 800 services, and these basic services continue today. These services were built on a network enabled by signalling system number 7 (SS#7) and IN continues to build on SS#7. More recently IN implementations cover a more extensive set of services from time of day routing plans, find-me follow-me services, pre-paid mobile services (wireless intelligent networks), calling card services, to advanced network-based call centre agent skill routing.

The basic aim of IN is to decouple the service logic from the control of the switch fabric, a stage further than could be achieved by the use of a stored program controller, and to create a platform that allows new services to be constructed from smaller building blocks. This later capability is expressed in Q.1201 as "integrated service creation and implementation by means of the modularised reusable network functions". The principle business aim of IN is the removal of a dependency on switch manufacturers for the provision of new services. In order to achieve this aim, the use of an SS#7 infrastructure is a prerequisite. Section 1.2 discusses the protocol stack of SS#7 and mentions the IN protocol – Intelligent Network Application Part (INAP). It is INAP that facilitates the decoupling of service logic from switching functions.

The development of IN service capabilities within the European Telecommunications Standards Institute (ETSI) is specified in releases called Capability Sets (CS), the first (obviously) being CS1. The ETSI work represents a clarification of the ITU-T standards and tries to add real-world experience to the theoretical standardisation effort. Each capability also creates a new release of the Intelligent Network Application Protocol (INAP) to support the remote invocation of services in an intelligent network, by providing the mapping from functional to physical entities (see later in this chapter). Further work has taken the IN standardisation to CS4, however, the actual number of implementations of anything other than CS1 are low[1]. In Europe a number of IN implementations are based on what has become known as ETSI core INAP – CS1, specified in ETS 300 374-1.

The time for IN is arguably gone, with new computing platforms and open standards for service execution environments just around the corner (see Chapter 10). The author also believes the telecommunications industry has become restless and disappointed with IN implementations and standards progress. The new software-based platforms called softswitches (see Section 10.3) being developed and session initiation protocol (SIP) servers (see Section 5.6) will replace the IN infrastructure; Ovum predicts this will happen by as early as 2003.

This sounds like a bleak future for IN, so why cover intelligent network technology here? The truth is that a lot of network operators have implemented IN successfully (and at considerable sunk cost), and are now sitting on a cash cow that will continue to deliver revenue beyond 2003. This revenue will be ploughed back into the development of new services in one of two ways, either to continue the evolution of their incumbent IN platforms to bridge circuit switched and packet switch worlds (in Chapter 10.4 the future of IN services is discussed and the work of a group in the Internet Engineering Task Force that looks to provide the ability for IN to interact with Internet-based services and vice versa), or to purchase softswitches and SIP proxy servers (see Chapter 5). Also the concepts for a distributed service network that are introduced in the IN standardisation are extremely useful in understanding the way new services in a call server architecture might be developed and delivered (we'll explore this further in the next section on services and explore how IN as it is

---

[1] It could be argued that CS3 implementations exist, because CS3 brings IN inline with the next generation of mobile networks under the standardisation for the Unified Mobile Telecommunications System standardisation (UMTS) activity and describes some of the IN services for mobile use.

now could evolve to have a place in the next-generation networks, see Section 10.4).

3.2 FUNCTIONAL COMPONENTS

The IN is decomposed into a number of layers, each of these layers are called planes, which form the IN conceptual model – INCM (ITU-T, Q.1201). This conceptual model is a framework for the construction of an IN architecture, not the architecture itself, an important distinction. The four planes are: service plane (Q.1202), global function plane (Q.1203), distributed function (Q.1204) plane and physical plane (Q.1205) (Figure 3.1).

I'll only examine the planes at a high level. If you would like to explore IN further than is presented here, I suggest [FAYN] as a good place to start, followed by [IAKO] for how IN-based broadband networks have been specified (the suspicion of the author is that the broadband IN standardisation effort will/has been overtaken by the adoption of application frameworks and protocols such as media gateway control protocol (MGCP) and SIP[2]).

The service plane is where the services reside. The global function plane describes a model network of functionality from a global or network-wide perspective. The distributed function plane describes, in a supplier independent way, the functions of the components that constitute an IN. The physical plane is where real network hardware resides and it embodies the functions described in the distributed function plane, and arguably is the easiest plane to understand.

The service plane recommendation (Q.1202) is a short document, four pages to be exact. It describes how a service can be categorised and describes the undesirable feature of service interaction. This is where the capabilities provided by one service if used in conjunction with another service produces an undesirable result. For example call waiting, a service designed to inform a caller that the person they're trying to reach is already engaged in a call, whilst also informing the person already on a call that someone else is trying to reach them. This service generally allows the called party to toggle between their existing call and the new caller to establish alternating conversations between the two. Additionally the service also offers the ability to connect all three parties together for a conference call. Clearly this service requires the called party to be able to control the calls they have both in progress and trying to reach them, it would be incompatible to try to implement this call waiting service at the same time as implementing a call diversion on busy service,

---

[2] See Chapters 12 and 5 respectively for an explanation of application frameworks, MGCP and SIP.

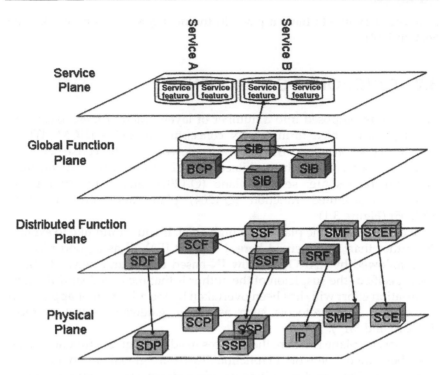

Figure 3.1 IN conceptual model

since call diversion on busy would try to divert the second caller away before any conversation could take place. Clearly, these two services can't coexist.

The global function plane describes the *proper* location for the modularity aspects of services construction. The concept of a Service Independent Building Block (SIB) is introduced. It describes Points of Initiation (POI), Points of Return (POR) and Basic Call Processing. The relationship of these to each other (in the form of the Global Service Logic (GSL)) is shown in Figure 3.2.

SIBs are standard (i.e. specified in the standards for each capability set) reusable functional components that services can be constructed from. Each SIB describes a single complete atomic capability that is independent of any physical component of the network. A service is created by effectively chaining SIBs together, these chains of SIBs create what is referred to as the GSL. The Basic Call Process (BCP) is in essence a special SIB that allows the passing of call related data from the other network elements, and in this way provides both a mechanism for handing over control of a call to the IN service and provides basic call processing functionality to the service logic. The BCP SIB defines entry and exit points to the call processing so that other SIBs can be invoked. These interaction points are

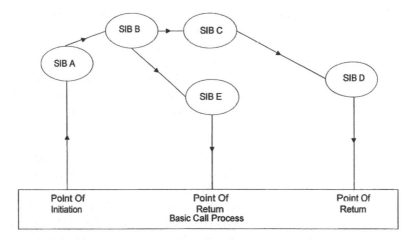

Figure 3.2 Relationship of SIBs, POI and POR to BCP

the POI, where call processing is handed over to the service (specifically the GSL), and POR where the results of the service execution hands back control to the call processing software.

The Distributed Function Plane (DFP) describes the functional elements/entities of:

- Service Management Function (SMF), the provisioning, billing and fault monitoring functions.
- Service Management Access Function (SMAF), interface between administrators and the SMF.
- Service Creation Environment Function (SCEF), the offline function for the creation of service logic from the SIBs.
- Service Control Function (SCF) controls the execution of services and contains a function known as the Service Logic Execution Environment (SLEE), where the combined SIBs in the form of Service Logic Programs (SLPs) execute.
- Call Control Function (CCF), basic call connection and release, essentially the inheritance from the stored program controller basic call control capability.
- Service Switching Function (SSF), the interface between the CCF and SCF manages the triggers in the CCF for interaction with the IN service. INAP is the protocol used to connect the SSF and SCF. This is in essence an instance of the special BCP SIB described above.
- Special Resource Function (SRF), provides such things as digit recognisers, recorded announcements and voice recognition capabilities for the interaction between callers and service logic.
- Call Control Agent Function (CCAF), essentially the signalling between the user and the network for example Integrated Services Digital Network (ISDN) Q.931.

The physical plane is populated with: Service Control Points (SCP, the physical embodiment of the SCF), Service Creation Environment (SCE, the physical embodiment of the SCEF), Service Switch Points (SSP, the physical embodiment of the SSF), Signalling Transfer Point (STP), Specialised Resource Platforms (SRP, also called Intelligent Peripherals (IPs), the physical embodiment of the SRF) and finally the Service Management Platform (SMP, the physical embodiment of the SMF and in some instances the SMAF too). Their relationship is shown in Figure 3.3.

The SSP is the stored program controller that we discussed earlier, in the case of IN the code for the call model (called the Basic Call State Model (BCSM), essentially the CCF) finite state machine has been modified (with the SSF) to include break points that allow the call routing to be paused for originating calls and terminating calls called the o_BCSM and t_BCSM. The break points in the SPC BCSM are called Detection Points (DPs). Each one of these triggers is separately settable either by a maintenance action or via a message from the SCP logic, to allow the call progress to be interrupted as various stages in the call processing (called Points In Call (PIC)). The PIC essentially map on to the POI and POR described in the BCP SIB defined in the global function plane.

The CS1 specifications do not allow for interaction between the originating and terminating call models and describes what are called type A services, i.e. single point of control (only one SCF can interact with the

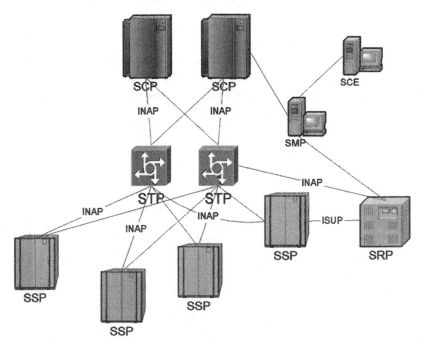

Figure 3.3 IN physical architecture

BCSM at a time) and single ended (SCF can only interact with an isolated half of a call, originating or terminating, but not both).

The STPs are responsible for routing SS#7 messages and whilst are present in the IN architecture, are actually part of the SS#7 architecture as discussed in Chapter 1. STPs were discussed in relation to a feature known as Global Title Translation (GTT). With reference to Figure 3.3, we will explain the importance of GTT. If the service code, say 0800 is used as a Global Title (GT) (i.e. the reference used to find an instance of a service that can be invoked) then the STP in the IN architecture can choose to route the INAP messages via Signalling Connection Control Part (SCCP) GTT to either of the SCPs in the diagram. This means that load sharing can take place and also if an SCP fails, then the failure would be transparent to the switch.

That's a whistle-stop tour of IN that hopefully gives a flavour of the work and thought that has gone into defining the way services can be created in a manufacturer independent way. The IN Conceptual Model (INCM) and the work of the TINA-C group (not described here) have gone a long way in assisting in the thinking process for Next-Generation Network (NGN) services and in some networks the current IN infrastructure may actually evolve to be the NGN service function or call agent capability (see Section 5.6 for a description of a call agent). In Part 2, Chapter 11, the IN's integration with call centres will be explored as a current service that can be improved upon by the use of NGN services.

4

Mobile Networks

4.1 INTRODUCTION

Mobile telephony networks have been a phenomenal success story since their introduction in the mid- to late 1980s. This success is being built upon, and a number of operators (old and new) around the world are looking to provide the latest generation of mobile network technology labelled 3G for third generation. The 3G label is based on the generally accepted premise that the first-generation cellular networks are the analogue-based ones first made popular in the late 1970s and early 1980s (AMPS in the US, TACS in the UK, NMT450/900 in Scandinavian countries and an NTT standard in Japan, more on these later), and that the second generation the digital cellular networks that arrived in the early 1990s (global system for mobile communications (GSM) in Europe, personal digital cellular in Japan, D-AMPS or IS-54, IS-136 and IS-95 in the US, again more on these later in the chapter).

The 3G mobile networks are subject to a set of standards developed by the International Telecommunications Union (ITU) (formally CCITT), Europe Telecommunications Standards Institute (ETSI) and the European RACE project. This work was started as long ago as 1986.

The ITU concept is based around handset mobility, and the early programme was dubbed future public land mobile telephone system (FPLMTS). The concepts were expanded to include the idea that a user should be able to access any telecommunications service from any suitable terminal connected at any point on any network. This became known as personal mobility. The ITU went on to define the concept as Universal Personal Telecommunications (UPT). The ITU was dragging its feet on what the standard should finally look like and the choice of technology for FPLMTS. FPLMTS was eventually renamed international mobile commu-

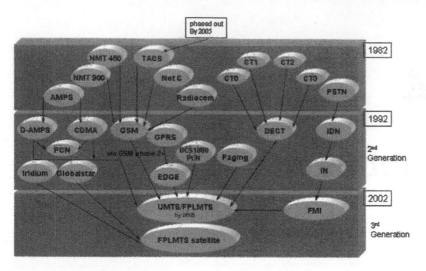

Figure 4.1 Roadmap to UMTS

nications for use in the year 2000 (IMT2000). Figure 4.1 shows the roadmap from first- through second- to third-generation network (IMT2000)
standards. This figure also shows a number of other technologies that
have been considered in the standardisation work as part of the change
towards the ultimate aim of IMT2000. These are a mixture of satellite
communications technologies (iridium and global star), paging technologies and cordless telephony standards of which Digital Enhanced Cordless Telephony (DECT) has seen the most recent and widespread use.

Work on advanced telecommunications services started in Europe even
before the CCITT (ITU-T) work. In 1985, the European Commission spon-
sored research and development in advanced communications technologies in Europe (RACE). This was initially intended to lead research
towards integrated broadband networks, but the work of one of the
projects considering the implications of radio-based mobile communications developed the idea of Universal Mobile Telecommunications System
(UMTS). This work floundered for a while, but was kick-started in 1995,
with the Bangemann report.

The recommendations that came out of this report produced the following roadmap:

- the regulatory framework for UMTS should be defined by the end of
  1997;
- basic UMTS working should be available by 2002;
- full bandwidth capability should be available by 2005;
- additional spectrum allocation in 2008;
- a UMTS forum should be set up to provide guidance for organisations
  such as ETSI.

ETSI have been influential over the last 2 years in delivering significant standardisation under the 3rd Generation Partnership Programme (3GPP) and have been delivering on the recommendations laid out in the Bangemann report. This work has also incorporated the ideas and work on IMT2000 and the reader can treat UMTS and IMT2000 as synonymous.

So why the need for change, why are all these technologies having to change? Simply put mobility has become a *'wanna have'* of modern society, it is trendy, affordable and practical to own a mobile 'phone, some might argue a necessity, as people take more control over their lives and jobs become more demanding, information and communication on the move is rapidly becoming the norm.

The GSM-based systems have gained enormous public subscription in both forms (GSM 900 and DCS-1800), not only in the UK, but also throughout Europe and the world. This has meant any change from GSM specification networks towards a so-called third generation would cause significant trauma. It is because of this, the move is seen to be evolutionary with a migration towards the new generation of networks and services.

It is also a given that a move away from the first-generation mobile systems, (TACS, AMPS, etc.) is a natural progression and to that end a move to free up the frequencies in the UK used for TACS and ETACS by 2005 has been undertaken. This combined with cost of maintaining old equipment is the reason why BT Cellnet (re-branded mmO2) pulled the plug on their analogue network at the end of 2000.

Under the 3G network systems, the aim is for users to be able to roam among countries that currently use different technologies and also for users to be capable of seamlessly moving between multiple networks, fixed and mobile, cordless and cellular. As a result, product longevity for core network and transmission components should be longer, and network operators should benefit from increased flexibility.

Universal Mobile Telecommunications System (UMTS) and more specifically 3G mobile systems have become infamous because of press about the licence auctions that a number of countries have run to sell off the potentially lucrative licences, only time will tell if these licences prove worth their money. That said the potential services offered by 3G networks are very exciting and look to have a potentially huge worldwide market. The change will not be easy as the mobile networks will undergo a similar change from a circuit switched infrastructure to a packet switched infrastructure as mobile telecos switch off their analogue services, and migrate their digital (GSM et al.) networks to a UMTS network.

4.2 MOBILE NETWORK ARCHITECTURE AND COMPONENTS

After that brief précis on the evolution and standardisation of the mobile networks, what are the actual components and how do they work.

All mobile phone systems are based on the cellular principle, which is a cluster of radio antennas arranged as cells all transmitting and receiving radio signals at different frequencies (Figure 4.2). The number of frequencies used can be reduced by allowing the reuse of the frequencies in cells that are sufficiently far apart to avoid interference.

What differentiates each system, analogue or cellular, is the way in which multiple voice signals are encoded between the handset and the radio station at the centre of the cell, called the base station.

Analogue Systems

In the UK, Total Access Communications System (TACS) is the analogue service offering started in the mid-1980s. It is a derivative of the American analogue system Advanced Mobile Phone System (AMPS). BT Cellnet and Vodafone originally offered TACS under the first mobile network licences granted in the UK. Capacity constraints led the UK government to release more radio frequency spectrum, around the 900 MHz band already used, to the cellular network operators and this created extended

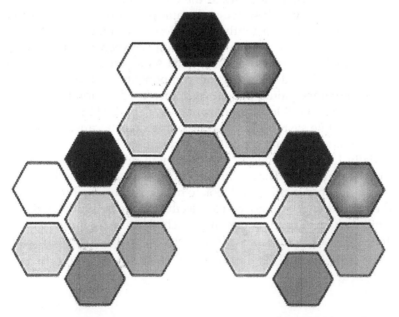

Figure 4.2 Repeating cell pattern colours indicate different frequencies

TACS (ETACS). Dual frequency handsets are pretty much the norm and users of the service perceive no difference.

In the US, AMPS was (and still is) the analogue service offered in most major cities from around 1983. The AMPS standards have been reused by both the UK and Japan as TACS and NTT's analogue systems both borrow ideas from AMPS.

In the Scandinavian countries (and Germany for that matter) Nordic Mobile Telephone systems NMT450 and NMT900 (operating at 450 and 900 MHz, respectively) were introduced in late 1981. This early release was taken up well, the population of the Scandinavian countries had a thirst for technology and Sweden in particular had at the time the largest penetration of both computers and telephones in the world.

Japan was actually the first in the world to release a cellular radio telephony system in Tokyo in 1979.

The AMPS-based systems use frequency division multiplexing of the voice channels to allow multiple users to access a particular cell base station. The network infrastructures are the same and are explained in the next section and shown in Figure 4.3.

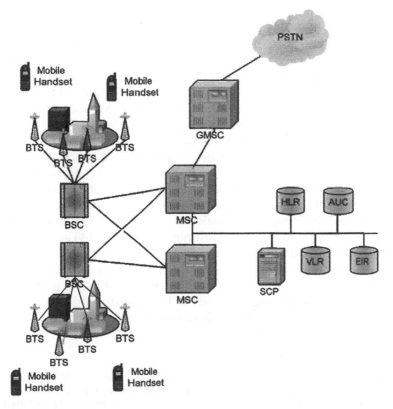

Figure 4.3   Mobile network architecture

Digital Systems

Global System for Mobile communications (GSM) is a European standardisation incentive with digital voice encoding on the air interface, developed as part of a standards effort to ensure compatibility for roaming subscribers. In the UK, Vodafone and BT Cellnet (now branded mmO$_2$) offer GSM services. As do One2One/T-Mobile and Orange, except strictly speaking they offer digital cellular system 1800 (DCS-1800). This is essentially GSM, but with an air interface operating at 1800 MHz. With a large number of roaming agreements with other network operators throughout the world GSM is arguably *the* global standard. It is this compatibility and roaming potential that makes GSM far superior to the first-generation analogue systems such as TACS.

GSM, IS-54 (D-AMPS), IS-95, IS-136 (IS stands for interim standard and are the standards applicable to US networks) and DCS-1800 networks all share a similar approach and topology to solve the issue of terminal mobility, and rely on the network architectures developed for analogue systems.

All mobile networks, like their fixed networked cousins, rely on the services of signalling system number 7 (SS#7). In the case of GSM, the addition of the Mobile Application Part (MAP) to the top layer of the SS#7 signalling stack enables the communication and control of mobility. In North America (and essentially all other countries that don't use GSM), Telecoms Industry Association (TIA) interim standard IS-41 finalised as the (American National Standards Institute) ANSI-41 standards for intersystem messaging is the equivalent in function to MAP. MAP and ANSI-41 sit at the same level in the protocol stack as the Intelligent Network Application Part (INAP) protocol (see Chapter 3).

The digital mobile network standards mainly differ around the specification of the radio interface; the US standards listed are in fact mostly specifications of the air interface. GSM has both digital voice (with time division multiple access) and digital (packetised) signalling. D-AMPS has digital voice encoding, with analogue signalling (AMPS signalling in fact). IS-95 uses Code Division Multiple Access (CDMA) as a mechanism for sharing the bandwidth to the base station. CDMA allows all the cellular phones to transmit at the same time, voice channels from different customers are separated in the base station by the use of a shared code value that allows the original voice to be reconstructed. It's like lots of people of different nationalities all speaking simultaneously to a partner in their national language.[1]

---

[1] Lots of battles have since raged over the use of CDMA for the future 3G networks, which were all financially driven around patents owned by Qualcom for the CDMA technology. These battles where eventually sorted out to everyone's satisfaction and hopefully for the better good of 3G networks.

The key components of a cellular network are:

- Mobile stations (handset plus smart card subscriber ID module).
- Base stations, the actual cellular masts (Base Station Transceivers (BTS)) and Base Station Controllers (BSCs).
- Mobile Switching Centres (MSCs), essentially the equivalent of a Public Switched Telephone Network (PSTN) service switching point (SSP). There is a variant of the MSC called a Gateway MSC (GMSC) which deals with the interconnect between the PSTN and the Public Land Mobile Network (PLMN).
- Mobility and management databases, Home Location Register (HLR) and Visitor Location Register (VLR). The HLR can also incorporate an Authentication Centre (AUC) or this is sometimes a separate database, to validate a handset on the network. An Equipment Identity Register (EIR) is also present in GSM. For the storage of the mobile station equipment identity.
- Where additional intelligent network services are used, an IN Service Control Point (SCP) is also present.

They are the main elements, Figure 4.3 shows how they fit together.

In order for calls to reach the mobile terminals, the network must know of their existence. When a mobile handset is turned on it initiates a registration and location update process. This process not only informs the network of the device's presence, but also uses the information in the EIR and AUC to ensure the device and user are valid, this process involves the VLR, HLR EIR and AUC. Once the device and subscriber details have been validated, the device is given a temporary roaming address (MSRN – Mobile Station Roaming Number). The reason for this temporary address is so that the network can constantly update the location of the device, whilst maintaining a fixed address to reach it, the fixed address being the mobile number called the MSISDN (Mobile Station ISDN Number).

Registration only takes place once, when the device is powered up, however, as the device moves around its home network it must constantly change its location information. All this information exchange is kept secret by the use of encryption performed both in the network and in the smart card Subscriber Identification Module (SIM).

Now if someone wishes to reach a mobile device, the address they use is the MSISDN. The MSISDN has no location-specific context; the only thing known is that a particular number range is allocated to a particular PLMN. If the call is originating from the PSTN the call will ingress through a GMSC. The GMSC will have to interrogate the HLR to discover the location of the handset which will invariably be roaming somewhere on the network. The HLR returns the MSRN, this number points the call to the MSC that is handling the group of BSCs that the mobile handset is on. The 'local' MSC then determines which BSC the handset is currently located with. The device is then paged by transmitting the paging message from all the MTSs that are in the mobile handset's location.

Clearly as a mobile device continues to move around a PLMN, procedures must be in place to handle the movement of the device from area to area and across different BSCs and even different parent MSCs.

This is a very simple description of the procedures necessary to call a mobile device in a GSM network. Other networks operate in a similar way. The main points to note here are that the PSTN only knows that a particular MSISDN number belongs to a specific PLMN if the owner of that number has ported their number to another PLMN, then the original PLMN must forward the call on to the new network. The alternative is to have a national database of all ported numbers that the PSTNs and PLMNs can access before routing calls. The point worthy of note is that a similar situation exists as the one previously described for customers that are roaming on a partner network (generally internationally). In order for the caller to be reached, the call is routed to their home PLMN. If you would like a more complete description of GSM, then I suggest you look up [EBERS]. All the transactions discussed above rely on the services of the Mobile Application Part (MAP) protocol (either GSM MAP of ANSI-41).

4.3   BEYOND GSM, THE PATH TO UMTS

GSM has undergone enhancements to its specifications with the aim to move it closer to UMTS. This work is being done by the 3GPP. This work is part of ETSI GSM and UMTS strategy and takes GSM networks through release 2 and 2.5. Release 2.5 introduces Customised Applications for Mobile Networks Enhanced Logic (CAMEL), General Packet Radio Service (GPRS) data services. Enhanced Data Service for GSM Evolution (EDGE) picks up where GPRS stops and forms the longer term data service for UMTS networks.

Enhanced Data Services

Phase 2 GSM data services are based on a single circuit switched connection capable of carrying around 9.6 kbps. This data rate is painfully slow for the future requirements of a multimedia service. Because of this, standardisation took place to improve the capabilities of the GSM network data services. This resulted in the specification of GPRS. A complimentary technology to GPRS is the High-Speed Circuit Switch Data service (HSCSD).

The main difference between GPRS and HSCSD is that HSCSD uses the existing Time Division Multiplex (TDM)/circuit switched infrastructure. This is how HSCSD operates. HSCSD uses the channels on the air interface to increase the data rate by bonding multiple channels together. If each TDM channel can handle 9.6 kbps, then up to six channels bonded together can give the same speed as most fixed line modems (56 kbps).

There is an upper limit of eight channels on the air interface imposed on HSCSD. However, the maximum data rate of a single TDM channel between the base station and the mobile switching centre is 64 kbps, so data rates higher than 56 kbps are likely not to be implemented. HSCSD has the characteristic of being a 'bit of a hog' on the air interface and for busy cells, users requesting rates higher than 19.2 kbps will probably be turned away.

GPRS allows enhancement to GSM to allow packet-based communications both over the air interface and through the core network. Like HSCSD GPRS uses multiple channels on the air interface so suffers the same problems for busy cells. This will allow for the use of Internet protocol (IP) datagrams from dual mode mobile handsets and integration to Internet service providers' equipment directly from the GSM network. The distinction being that phase 2 GSM data services are based on circuit switched connections with modems, just like the telephone network (except restricted to around 10 kbps).

GPRS is an overlay network on the existing GSM network and involves the introduction of two additional key nodes the Gateway GPRS Support Node (GGSN) and the Serving GPRS Support Node (SGSN) (Figure 4.4). The SGSN performs the following functions:

- authentication and authorisation
- admission control
- usage data collection for billing
- packet routing
- mobility and link management

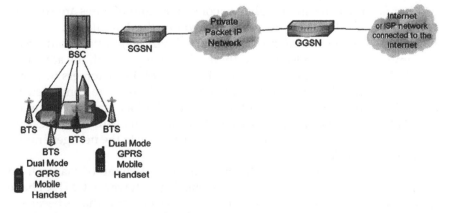

Figure 4.4   GPRS overlay network

The GGSN is the gateway out to the public data network in most cases this is the Internet and/or the mobile operator's Internet Service Provider (ISP) network infrastructure. In order to make use of GPRS new handsets will be required.

Beyond GPRS we have Enhanced Data service for GSM Evolution (EDGE). EDGE improves upon GPRS (and HSCSD) by modifications to the way the signal from the handset to the base station is modulated.

Mobile Intelligent Networks

CAMEL introduces the intelligent network concepts to GSM and introduces the idea of a Virtual Home Environment (VHE). CAMEL makes network services transparent of location through the introduction of IN style services. In order for CAMEL service to be present in a mobile network, the mobile network architecture will most likely include an intelligent network Service Control Point (SCP).

The most common example service quoted is that of reaching a voicemail service. Most networks offer a short code to access a voice mailbox (for example Vodafone in the UK offers 121). When a customer is roaming in a partner PLMN these short codes are not available, CAMEL addresses this issue. This is achieved through signalling all incoming and outgoing calls from a CAMEL subscriber through the CAMEL Service Environment (CSE) using CAMEL Application Protocol (CAP). The foreign network's VLR must obtain the CAMEL Subscription Information (CSI), which informs the foreign network of the CAMEL services the subscriber has. CAMEL is the wireless intelligent network standard for GSM networks. The equivalent is WIN (Wireless Intelligent Network) for ANSI-41-based networks.

If you want to know more about GPRS and CAMEL then check out [SIEG].

Other factors including regulatory constraints and dictates are shaping the move towards UMTS for example within the UK market the four players (BT Cellnet, Vodafone, Orange and One2One) initially had specific numbering ranges allocated to them. This changed with the advent of Mobile Number Portability (MNP), allowing mobile customers to retain their mobile number irrespective of the network they are connected to. The MNP directive issued by Oftel (in the UK) has featured in pretty much all the mobile marketplaces, both as a mechanism for increasing competition and also as a means of easing the move to the next generation of mobile networks.

Figure 4.1 earlier in this chapter, indicates the roadmap of change from cellular telephony and other wireless technologies towards UMTS. I have only covered GSM in any detail in this short section and that is because I am most familiar with this technology. The North American standards are equally important and present their own set of challenges if a move to a global universal mobile service is to be adopted. If you would like to know more about the US mobile technologies and services I suggest you refer to [BLACK3].

In closing, one final note on IN, circuit switched technologies and mobile communications in the form of GSM, IS-95, IS-136 and other mobile technologies, they are all major contributors to the design of new services. These new services will evolve through techniques such as: the Wireless Intelligent Network (WIN), the combination of IN and mobile technologies to provide services such as pre-paid mobile; the evolution of the standardisation of CAMEL; the inclusion of new platforms such as SIP proxy servers (see Section 5.6). GSM, UMTS and WIN are books in themselves and deserve more than this short introduction alludes to. I recommend the reader consult [CHRIS] and [EBERS] for more information on WIN and GSM.

Third-generation networks include the use of packetised voice carried over IP and work is also underway to separate the functions of the MSCs into softswitches and media gateways (see Sections 5.6 and 10.3). The core processing functions of the MSC are also looking to be standardised under the Open Service Access project, to provide an open platform for supplier independent service creation (see Section 12.3 on application frameworks).

Third-generation mobile networks will become an important part of lots of people's lives in the 21st century, in Chapter 13 a scenario is described that will be commonplace in just a few years time, ubiquitous access to multimedia services.

5

Packet Switched Technologies

5.1 INTRODUCTION

In the same way we started with the transmission infrastructure for circuit switched technologies, packet switching has a similar underlying infrastructure. At this stage, it is worth reviewing the International Standards Organisation's Open Systems Interconnect model (ISO-OSI) that was used in Chapter 1 as a framework for the telephony signalling protocols (Figure 5.1).

In the case of packet-based networks such as the Internet, since this is the one of specific interest to us, the bottom two layers of the OSI stack are populated with technologies that differ based on their geographic boundary: local area, campus or metropolitan area and wide area.

In Local Area Networks (LANs), the infrastructure is now predominantly Ethernet (in its numerous forms of 10 Mbps, 100 Mbps and 1 Gbps). Ethernet is a general term used to cover the standards developed by the Institute of Electrical and Electronic Engineers (IEEE) under the number 802.3. 802.3 is based on the work by Xerox, who coined the name Ethernet, based on the term *luminiferous ether*, through which Victorian scientists first thought electromagnetic radiation travelled.

Token ring is another LAN technology originally developed by IBM and adopted by the IEEE as 802.5. The idea is to timeshare access to the network by the use of a token that a device (say, a PC) must acquire before it can transmit data on the network. Token ring, also had a significant number of installations, but it is probably safe to say has largely been usurped by Ethernet.

In the campus or Metropolitan Area Network (MAN) technologies such

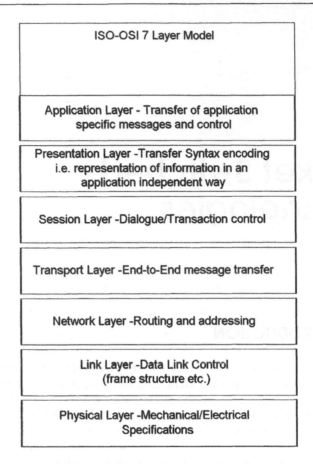

Figure 5.1   ISO seven-layer model

as Switched Multimegabit Data Services (SMDS) and Fibre Distributed
Data Interface (FDDI) are used for providing backbone connectivity
between LANs. SMDS is a connectionless high-speed LAN interconnect
technology (that has not been widely taken up). FDDI is a token-based
technology (not dissimilar to token ring) running at 100 Mbps and has a
ring circumference of up to 200 km. FDDI has been extensively used to
connect computing equipment together in large server installations, but
has been largely replaced by gigabit Ethernet.

More recently Asynchronous Transfer Mode (ATM, see Chapter 7) and
gigabit Ethernet have gained prominence in campus networks for high-
speed LAN interconnect. Whilst time division multiplexed leased circuits
together with frame relay and X.25 has remained the predominant technology for wide area interconnects, ATM has gathered ground for wide
area interconnects, but still remains a minority compared to leased
circuits. Gigabit Ethernet may yet be the candidate for the throne of

Wide Area Network (WAN) connections, with international gigabit Ethernet now being offered.[1]

Whilst the technologies above proliferate at the physical and link layers, they will not be covered in any more detail here (the keen reader might like to read [TANE] for more detail). This is not meant to undermine the importance of any of the technologies mentioned, but to allow more focus on the network and transport layers of the OSI model.

A number of technologies have proliferated at the network layer: NetBEUI, IPX/SPX, AppleTalk and TCP/IP.

NetBEUI (NetBIOS extended user interface) is a protocol with its origins in IBM's NetBIOS (network basic input/output system) and was part of their PC network LAN product. The NetBIOS work was later adopted and extended by Microsoft as part of the Windows™ operating systems, to facilitate file and print sharing between peer computers, to form NetBEUI. NetBEUI has limited use outside the local area network as it has a fairly limited naming and addressing capability. Microsoft has really moved on from NetBEUI and has implemented file and print services over TCP/IP (called NetBT or NBT for NetBIOS over TCP/IP).

Internetwork Packet Exchange/Sequenced Packet Exchange (IPX/SPX) was an invention of Novell. They are network and transport layer protocols, respectively. They are used to support the Novell NetWare™ product. Their use outside of Novell NetWare networks is of limited interest, so they will not be expanded upon.

AppleTalk is the native protocol of the Apple Macintosh network connectivity used to facility file and printer sharing in Macintosh installations, and arguably has been replaced by TCP/IP.

Transport Control Protocol/Internet Protocol (TCP/IP) are the transport and network protocols developed within the Internet Engineering Task Force (IETF). The industry has firmly fixed on IP as the protocol of choice for both the local and wide area transport of data. It is the TCP/IP protocols and the application protocols above them that we will concentrate on in this chapter.

Whilst IP has gained dominance as the network layer protocol of choice, what hasn't occurred is the choice of a clear winner for the link and physical layers. This may be about to change however, as work is underway to place IP directly on top of Wavelength Division Multiplexed (WDM) fibres. Maybe in the not too distant future as fibre is rolled out to more and more homes (economics aside!) individuals may get gigabit per second access with IP over a single (or multiple) wavelength WDM.

TCP/IP's dominance of the market means that all effort has been focused on delivering services over TCP/IP and placing TCP/IP on top of different link layer technologies such as the recently introduced Digital

---

[1] Telecommunications International Magazine December 2000.

Subscriber Line (DSL) services and cable TV Internet services (see Chapter 6 for more details on these).

5.2   BASIC INTERNET PROTOCOL

If you've read books on Internet Protocol (IP) then you can skip this section. I'll cover the areas of IP version four and a little on the importance, need and basic differences of IP version six. As indicated in the introduction to this chapter, IP has gained dominance as the network layer protocol.

In the section on signalling system number 7 (SS#7) (Chapter 1), we explored the network layer protocol called Message Transfer Part (MTP). The design aims of MTP were very different to that of IP. Whilst MTP is designed to facilitate the reliable transport of signalling packets across potentially unreliable connections via retransmission and checksums, IP on the other hand was designed with one goal in mind. To efficiently package packets and deliver them (route) through the network using addresses in the header. IP provides an unreliable, connectionless data delivery service.

What this means is that if packets get discarded (say a device handling the IP data runs out of buffers or crashes), then that's tough. The best IP offers is to inform the application using the IP layer that something went wrong. The same goes for sequencing of whole packets. If two packets are sent between the same source and destination, they can take very different routes between those endpoints. This means the second packet of data could arrive before the first. IP doesn't fix this it just delivers the packets in the sequence they arrive.

One exception to the rule on sequenced delivery is concerned with *fragmentation*. One of the properties of IP is to be able to break packets up into smaller *fragments* (for example a frame size restriction of the data link layer may prevent the whole message fitting into a single frame). Fragmentation causes the original packet to be broken up into a number of smaller packets. Each packet is then a separate entity from the others and can travel by a different route. IP will attempt to reassemble these fragments back into the original packet. However, if one of the fragments gets lost, the whole original packet must be discarded, and as we've already stated lost packets cause an error but it is up to the application or protocol above IP to take remedial action.

IP was designed in unison with Transmission Control Protocol (TCP). TCP/IP to give it the correct name (or Internet Protocol suite) is heavily reliant on other protocols such as Ethernet to carry the packets it encapsulates over the physical media. This is again unlike the protocol suite of SS#7 where the MTP layers go all the way to defining the data link and physical layer too.

Why is it important to highlight these differences? Because this book is about the move from the circuit switched environment of SS#7 to the packet-based world of IP and it is the differences between the two that make the transition all the more challenging. This challenge has been taken up by a working group in the IETF, called the SIGnalling TRANsport (SIGTRAN) group. We will discuss the work of the SIGTRAN group later (Section 5.6). It is sufficient to say for now, that this group has been focused on getting the MTP style reliability into carrying circuit switched signalling protocols on IP networks.

Enough on the comparisons with SS#7, now for some detail on IP. The thought that most people have is to ask the question "where did the Internet come from?" It seemed to appear very quickly and dominate. I guess like all technologies that succeed they seem to appear spontaneously and quickly expand. The truth of the Internet and packet switching in particular is that it has actually been around for a long time, 30 years (or more). The reality is that email actually first appeared in 1972. TCP/IP was first demonstrated in 1973 and became almost universally adopted in academic circles by 1983. In truth the rise of the Internet has taken some time, what really drove it from academia to mainstream is Marc Andreeson's invention of the graphical web browser in the early 1990s that utilised the protocols and ideas developed by Tim Berners-Lee.

So IP, it was invented nearly 30 years ago and is changing the world, what actually is it? All packet switched protocols fulfil the same role; they are a mechanism for enclosing the actual data to be carried in a wrapper. The wrapper generally contains a header and sometimes a footer or trailer to mark the start and end of the enclosed data. In the case of IP, only a header is present and the need for the trailer is obviated by the use of a count of the size of the packet. The header in IP also contains the next most important information, the source and the destination of the data. There are other fields in the header for the version of the IP protocol being used, and which higher level transport protocol to pass the contents of the packet to, the most obvious one being TCP,[2] amongst other items. One last important detail about the header, it is a variable length a minimum of five 32-bit words and a maximum of fifteen 32-bit words (including an area in the header called options), or 60 bytes. The field that counts the size of a packet is 16-bits long allowing a packet (including header) to be a total of 65,535 bytes long. This maximum is currently rarely reached for example, for 10 Mbps Ethernet the maximum frame size allowed is 1500 bytes. This means that very large packets would have to be broken up (fragmented) into smaller chunks. Fragmentation costs in processing time and thus introduces latency (for voice we've already covered the fact that latency is not a good thing, more on this later).

---

[2] UDP is possibly the next one to be thought of, but lots of others exist also, these are listed in RFC 1700.

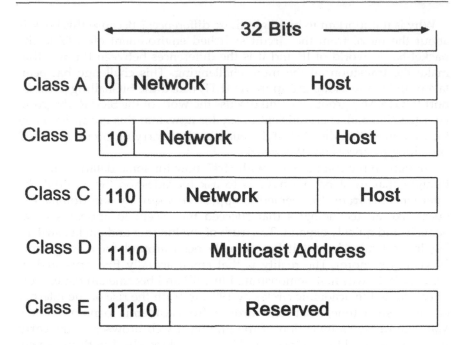

Figure 5.2 IPv4 address formats

The rise of higher speed networks such as gigabit Ethernet may start putting pressure on this 64k limit on packet size (gigabit Ethernet (802.3 Hz) has a maximum frame size of 9000 bytes), so the previous statement may not hold true for long.

The address fields in the header contain a 32-bit address each (that's 4 billion possible addresses); everything on the Internet has a unique address.[3] The address field is split into a network and a host portion. The class of the address dictates the number of networks and the number of hosts per network (see Figure 5.2). The classes are A, B, C, D and E, this is purely a convenient means of grouping addresses into usable blocks and to assist in routing.

IP addresses are normally written in what is called '*dotted decimal notation*'. Each 8-bit portion of the 32-bit address is written separated by periods. For example the 32-bit address C245CDB3(H) is C2.45.CD.B3 and in dotted decimal 194.69.205.179 and this is in fact a class C address. How can you tell it's a class C address? Looking at the left most portion of the address C2, the top 1–5 bits based on their pattern determine the class.

- for class A addresses the top most bit is 0

---

[3] This note on unique addresses is actually only true in the context were addresses are not translated in any way. When IP addresses were thought to be becoming scarce, a technique called Network Address Translation (NAT) was created, see later in this chapter.

- for class B the top most 2 bits are 10
- for class C the top most 3 bits are 110
- for class D the top most 4 bits are 1110
- and finally for class E addresses the top most 5 bits are 11110.

So we can see that Hex C2 represents 11000010 in binary and the top 3 bits are 110, hence class C.

The different classifications each split the address space down into different numbers of networks and hosts:

- class A has 126 networks with up to 16 million hosts
- class B has 16,382 networks and up to 64,000 hosts
- class C has 2 million networks and up to 256 hosts per network
- class D is a special address range reserved for multicast, in which a packet can be sent to a single address but received by multiple hosts
- class E addresses are notionally reserved for future use (what ever that might be!)

In order to make good use of an allocated address range, for example a class B address range 191.10.x.x, which represents one network with up to 64,000 hosts. If you require more than one network segment say in two offices with 200 PCs each, should you apply for another IP address range? That's one solution, the other is 'subnetting'. A class B address has a 16-bit host address. Subnetting allows the network part of the address to be extended using a subnetwork mask. If the upper 8 bits of the host portion of the address where masked to form an extension to the network portion of the address, then to the outside world the network number will not have changed, network 191.10.x.x is still the allocated network. What has changed internally is that hosts with addresses of 191.10.1.x are now on what is referred to as subnetwork 1. And there are up to 254 hosts on subnetworks 1 through 254. What you aren't allowed is address 191.10.255.255 as a host address as this is the broadcast address for network 10.6.x.x. The broadcast address is a special address that means the packet is to be received by all hosts on that network.

What is the purpose of all this address segmentation? Quite simply 'reachability' is the answer. In order for the Internet to prove useful across a wide area, hosts need to be reachable, i.e. a packet from one host must be able to reach another host over a number of interconnected networks.

In the Internet, reachability is performed by the routing function. What is routing? Routing is the ability of the network to take the address of the destination host and forward the packets across multiple devices (routers) on to its final destination. Routing relies on the segmentation of the address space in order to scale in an acceptable way. Routing tables (the function that maps network addresses to ports on a device) hold pointers to networks, generally not hosts.

In order to maintain routing information in all the routers in a network, a number of additional protocols must be used: Open Shortest Path First (OSPF); Routing Information Protocol (RIP); Internet Control Message Protocol (ICMP); Interior Gateway Routing Protocol (IGRP); Border Gateway Protocol (BGP) and Address Resolution Protocol (ARP). Routing protocols are used to transfer routing information about network addresses and links to them.

ICMP is used by routers to signal the fact that something has gone wrong. It does this by sending messages to hosts and routers about the packets it receives and about the events it detects. For example one of the most common messages for applications to see is 'destination unreachable'. This means a router cannot determine where a host is located. Other common messages are the echo request and echo reply messages. These messages are manifested in the 'ping' application used to test reachability of the selected host IP address and whether it is 'alive'.

IP addresses are fine for routing on the broader network scale, but clearly there is a need to associate an IP address with a physical device (host) connected say to an Ethernet network. The Address Resolution Protocol (ARP) is used for this. ARP sends a broadcast packet out on the LAN to ask who owns a specific IP address. This message only has local context and all the hosts on that network see the ARP request. The host that owns the IP address that the request relates to replies.

Open Shortest Path First (OSPF), is a routing mechanism that takes into account three different parameters to control its routing decisions: delay, throughput and reliability. In order to make decisions based on these parameters, routers exchange messages (called link state updates) that tell the other routers in the Internet of the status of a particular route and a value for the parameters above. The value of the parameters is used to build a graph (topological view) of the network of routers that form what are called *adjacent* nodes in OSPF. The designation of adjacency is not proximity, but is based on the nomination of a specific router called the *designated router*, that all the other routers exchange information with. OSPF is a form of IGRP. The converse to interior routing protocols is obviously an Exterior Gateway Routing Protocol (EGRP). RIP was replaced by OSPF.

BGP is an exterior routing protocol. What is the purpose of the difference between interior and exterior routing protocols? The need to control the routing of packets between neighbouring areas owned by different businesses (such as different networks owned by different Internet Service Providers (ISPs)) is the reason for border routing protocols. A border gateway protocol like BGP allows policies such as don't route Oracle packets via a Microsoft-owned network. This allows the construction of the Internet from a group of separately managed privately owned networks (essential for an internetwork to span the globe). These separately managed networks are referred to as Autonomous Systems (AS).

When are we expected to run out of IPv4 addresses? That's an interesting question, Christian Huitema in his book on IPv6 [HUIT] gives estimates from work done during early work on IPv6 and a date somewhere between 2005 and 2015. A number of factors are potentially pushing this date towards the later time, more efficient use of addresses in routers (per router addressing, rather than per port) and most notable Network Address Translation (NAT). NAT is a mapping technique that can map a number of *'private'* addresses to a single IP address.

NAT was originally created as a technique for preserving the IPv4 address space. NAT has become a very useful security technique. It is now commonly deployed for this reason, rather than for address space preservation and can be commonly found in Integrated Services Digital Network (ISDN) and Digital Subscriber Line (DSL) access devices (see Chapter 6). The one major factor that could cause the depletion of addresses to occur earlier, rather than later, is the emergence of the mobile Internet in the form of Wireless Application Protocol (WAP) and the Japanese i-mode technologies (see Chapter 8) initially and with the evolution to third-generation (3G) mobile networks (see Chapter 4).

Other security measures have been employed (non-NAT) extensively in recent times, as more people have become 'connected'. The most common term used in IP security is firewalls. Firewalls are generally now specialist devices that incorporate two functions: packet filtering and application proxying. Packet filtering is a technique that uses a look-up table as part of the routing function described above, to selectively allow or deny packet forwarding to take place (thus denying access to specific destinations). Application proxying is more complex and relies on an application looking at the contents of the packets passed to the firewall and applying intelligence about what is contained in the packets based on the application the packets relate to, to selectively allow or deny the forwarding of packets.

IPv6 is the next release of the IP, designed to overcome issues of IPv4 and looking to the future when even more devices will be connected together. The first question that is always asked is what happened to version five. Version five was allocated to an experimental stream protocol so couldn't be used!

A lot of debate took place to create IPv6, but finally in 1994 a recommendation was published that formed the basis for IPv6 going forwards (RFC 1719). The discussions that took place are documented in [BRAD]. As is always the case, decisions are based on consensus, IPv6 is no exception and surprisingly the consensus was quite large (by all accounts).

The main areas of change are in: the size of the address space (128 bit); support for security, multicast, auto configuration and support for real-time communications.

That's the brief précis of IP. Clearly there is a lot more detail than can be covered here. The definitive guide has to be [STEV], so I refer you to this complete work.

5.3 MOBILE IP

In the increasingly sophisticated world of palm-based devices and personal digital assistants, the problem of terminal mobility presents itself. Terminal mobility is the property exhibited by any device that is portable, mobile handsets being the most obvious incarnation of a mobile terminal to date. Mobility in Global System for Mobile communications (GSM) is handled by the combination Home Location Register (HLR) and Visitor Location Registers (VLR) and a sophisticated set of signalling messages between the mobile device and the network (see previous section on circuit switched technologies). Mobile IP performs a similar role in tracking mobile devices and forwarding packets to them.

Routing (based on network addresses) is clearly not possible when the mobile device is potentially moving from network to network. When a device is not actively being used for communications, then arguably the IP address could be changed via say Dynamic Host Configuration Protocol (DHCP),[4] and then updated to a domain name server (see Chapter 9 on directories later for more on Domain Name System (DNS)). When a device is actively transmitting data, for example a real-time stream carrying voice, changing the IP address mid-session just isn't viable. Therefore, another means of forwarding packets must be used. This type of problem presents itself most obviously in new 3G mobile networks and a number of approaches have been proposed to support an Edge Mobility Architecture (EMA) in these cases [BTTECH], which include mobile IP as a component.

So, now for some more on mobile IP, mobile IP defines three main components that communicate via mobility protocol to create a domain where devices can roam whilst maintaining contact with each other.

- *Mobile Entity* (ME) or node – which is the roaming device that needs to maintain communication whilst it is roaming by virtue of a 'care-of' address. When in their home network, these devices operate as any other node on that network.
- *Home Agent* (HA) – a router with a connection to the ME's 'home' network. The word router here means a device, which forwards packets not destined for itself. The agent (since that's generically what agents do) acts on behalf of the ME to provide other devices wanting to reach the ME with a fixed point in the network to communicate with. The ME needs to keep the HA up to date with its current location (care-of address) at all times, so that the HA can forward packets to it. The HA intercepts packets destined for the ME and 'tunnels' packets to the ME. The tunnelling process involves placing the origi-

---

[4] DHCP is a protocol used to dynamically configure hosts with a range of network related information: IP address default router, name server addresses, etc.

nal packet into another packet with the ME's current destination. The destination could actually be the FA below.

- *Foreign Agent* (FA) – a router on a ME roaming (foreign) network, that assists the ME in communication with the HA. In some instances also de-tunnels the packets forward by the HA and passes them to the ME. Finally the FA is the ME's default gateway out to the wider internet-work.

In order to facilitate mobility, these entities need to communicate in the same way GSM relies on the handset periodically broadcasting its presence. HAs and FAs broadcast their presence on a network either via multicast or local broadcast packets. These packets contain 'advertise-ment' messages that MEs listen to and use to work out where they are – roaming or at home.

The advertisement packets that the FAs broadcast contain foreign addresses that are available to roaming nodes. The roaming node (ME) can acquire a foreign address using for example DHCP. Once the care-of address has been acquired, the ME informs its HA of this address.

The HA then is responsible for constructing a tunnel to the ME. The tunnel can terminate either at the FA or the ME itself. Packets in the other direction (back to the other node from the ME) are sent directly to the other node, with the source address set to the ME's Home Agent's (HA's) fixed address.

This represents a very high level view of mobile IP, but hopefully it demonstrates its usefulness in the next generation of data centric networks. If you want or need to know more then I suggest you consult the RFCs (2002, 2003, 2004, 2005, 2006 and 1701 for generic routing encap-sulation) or [SOLO].

5.4 TRANSMISSION CONTROL PROTOCOL

This section briefly looks at one of the two most commonly used applica-tion protocols that sit on top of IP, namely, Transmission Control Protocol (TCP), the other being the topic of the next subsection UDP.

TCP provides applications with a reliable connection oriented byte-stream service. In order for two applications to communicate with each other, the connection first has to be established, we mentioned broad-cast and multicast in the previous chapter on IP, these communications methods don't apply to TCP, there are just two entities, a client and a server. The client creates/initiates the connection and the server accepts it.

Reliability is provided in TCP in a number of ways: retransmission, confirmation (acknowledgement), checksums and duplicate segment removal and segment reassembly. Segments are what TCP sends, these are chunks of application data, the size of which TCP determines, not the application program. Additionally TCP manages transmission of information in a variable quality network through the use of flow control. Flow control stops buffer overruns and network congestion.

How does IP distinguish TCP from any other protocol? The answer is via protocol numbers. Every protocol (TCP being no exception) is allocated a specific, well-known protocol number.

How does TCP distinguish between different applications, so that when information is handed to TCP from the IP layer, it can pass it on to the correct application? Port numbers perform this function. A number of well-known applications have well-known ports, for example: SMTP (email transfer protocol) uses port 25, POP3 (email post office protocol) uses port number 110, FTP for file transfer utilises port 21 and HTTP (web page retrieval) uses port 80. The choice of port number outside the well-known ones is up to the application, in the case of a client program, it can just ask the TCP layer to allocate one automatically. For server applications that have to 'listen' to a specific port, then the choice is again open, however, port numbers below 1024 are generally reserved for use by well-known applications or Unix or MS Windows services. TCP port numbers and IP addresses uniquely identify a single bi-directional connection between two hosts. This four-tuple (source port, destination port, source IP address and destination IP address) is called a 'socket pair'.

Flow control in TCP is provided by what is referred to as a sliding window mechanism. TCP controls flow by exchanging a value of the number of bytes each end is willing to accept before an acknowledgement is sent, this is the window, as bytes are acknowledged, so the *window* slides along the bit stream until all the bytes are transmitted.

5.5 USER DATAGRAM PROTOCOL

User Datagram Protocol (UDP) as the name suggests is a datagram protocol, every chunk of data passed to the UDP layer from the application is packaged up and transmitted as is. This basic service clearly requires very little by the way of header information and this is in fact true; UDP has only 8 bytes of header information. UDP (unlike TCP) provides no reliability or payload fragmentation and is connectionless. Therefore, the application needs to be aware of the maximum size the IP packets can grow to (MTU), so that it can cater for fragmentation, this is because if IP detects a loss of a fragment in a datagram it must discard the whole datagram. TCP can cater for the discard by retransmission, UDP on the other hand has no such capability, and therefore the application using UDP must be aware of the discarded datagram.

UDP utilises port numbers in the same way as TCP, for the recognition of which application to pass the data on to. UDP has the ability to use checksums to ensure the data and header have been received from IP correctly, however, unlike TCP this checksum is not mandatory. If a checksum is used and the receiver detects an error, nothing happens except the datagram is discarded, no retries, no errors!

This may sound like UDP is not much use to any application. Actually UDP is useful for the transmission of packetised voice samples. Voice is not tolerant of latency, TCP with its retry mechanism can introduce great wads of latency in between datagrams, UDP on the other hand just doesn't worry about discards it doesn't try to retransmit it just ignores it. This works just fine for voice which is much more tolerant of the odd lost sample, silence can be introduced or white noise can be used to compensate the listener. This brings us nicely on to the next section multimedia transport. Just one final note before we venture away from UDP. UDP by virtue of the fact it is connectionless means it can utilise some very useful aspects of IP multimedia communications, broadcast and multicast. This makes UDP extremely attractive for all sorts of group communications services. We'll discuss more of these services in the next section of the book.

5.6 MULTIMEDIA TRANSPORT

Introduction

In the chapters on circuit, switched telephony we discovered that voice encoded at 64 kbps is carried on individual timeslots that make up a much larger multiplexed stream of calls that are transported and switched across the time division multiplexed network, being controlled by signalling protocols such as ISDN user part (ISUP) over MTP. This section explores how voice (and video) is transported and controlled across a packet switched network utilising IP as the base protocol.

A lot of the original work on voice and video transport over IP networks grew out of the research on the MBONE, a multimedia backbone network constructed as an overlay to the Internet. The MBONE utilised software written to allow multicast packets to traverse unicast tunnels across the Internet. The software managed the membership of multicast groups and the 'pruning' of routing trees needed to efficiently distribute multicast traffic in what was a predominantly unicast world of routers. The author was very fortunate to be reading for a master's degree in data communications networks and distributed systems at University College London, when a number of researchers were working on the MBONE, receiving some early exposure to the potential of IP-based telephony.

Real-time Transport (Control) Protocol

The Real-time Transport Protocol (RTP) is what is used as the mechanism for transporting real-time media such as voice over an IP network. RTP utilised UDP as the end-to-end transport mechanism on top of IP.

The main point about RTP is it in itself has no connection creation facilities or mechanisms for supporting Quality of Service (QoS), these functions are left to other protocols. Remember UDP is connectionless, so the actual media stream is not connected in any real way, unlike the circuit switched network. Also because UDP supports multicast so does RTP. RTP has to be able to cope with mixing of media streams for example when using RTP for a group conference.

RTP relies on the Real-time Transport Control Protocol (RTCP) for the end-to-end monitoring of the media stream. Both RTP and RTCP utilise UDP ports, RTP using even port numbers and RTCP the next highest odd port numbers by convention.

The RTP header is 12 bytes long and consists of:

- A number of flags (version, padding, extension to allow for addition header extensions, count of contributing source identifiers – to allow for mixing, a marker to for example mark frame boundaries).
- Followed by a payload type field which indicates the format of the contents for example 64 kbps Pulse Code Modulation (PCM) μ-law encoded voice, a number of different payload types are predefined in a profile. Once the stream is started, the payload type cannot be changed, and for that matter if the mixing capability is used and multiple sources are contained in the payload, then they all have to be of the same media type since there is only one payload type field.
- The sequence number field comes next; remember back to the description of UDP, the application needs to be aware of discarded datagrams (the datagram could of course just be delayed, not lost, but the effect is the same). That's what this field is for, as well as the fact that datagrams could arrive out of sequence, since IP doesn't give any guarantees over sequence. The sequence number field could be said to be superfluous as trying to reorder or even consider retransmitting a real-time voice or video sample would cost in the form of latency, something as we've already discussed is not good for interactive communications. However, for near real-time applications such as video or voice/music streaming at high quality where buffering is used, the sequence number would have a use to increase quality without dropouts.
- Next comes the timestamp field, this field contains timestamps relating to the media type and is used to determine network performance and to provide for RTCP to perform latency and jitter feedback.
- The Synchronisation Source field (SSRC) identifies the sender of the

payload. The relationship of this field to the Contributing Source field (CSRC), is that if there is only one sender and no mixing taking place then the CSRC which follows the SSRC field is set to zero.

• The CSRC field can contain up to 15 contributing source identifiers (count flag is 4 bits long). This field is only present if an RTP mixer is involved, the source identifiers are the original identifiers that have been inserted in the header by the mixer.

The companion protocol to RTP is RTCP, its purpose is to inform the endpoints of the RTP stream about quantities such as: packet delay, jitter and packet loss. Since RTP is connectionless, RTCP performs the feedback function that allows RTP to determine if packets were actually arriving at their destination, without this RTP would have no means of determining if anything was happening. Whilst RTCP allows a level of control over RTP it is important to realise the distinction of RTCP from a signalling protocol such as those described later, signalling is about the creating and termination of sessions, RTCP's purpose is to monitor the progress of the real-time stream. RTCP provides the information about the status of the stream through the use of five different packet types: a Sender Report (SR), a Receiver Report (RR), a Source Description (SDES), a BYE and finally an Application Specific Packet (APP).

Sender reports are from active RTP senders, receiver reports are sent from non-active participants, i.e. those that are only receiving RTP streams. The source descriptor packets allow a relationship between the SSRC value in the RTP header with a more real-world item such as an email address or name or both. Generally the SDES packets are sent at the start of a session so that participants can be identified. The BYE packet informs other participants that the sender is about to leave the session and terminate their RTP stream; also BYEs are used to indicate that a loop in a media stream has been detected and should be terminated.

Application-specific packets are pretty much the odd ones out. They are intended for the transmission of application-specific information, they were proposed for experimental purposes.

Before we move on to signalling protocol for IP networks, just some closing observations of the use of RTP over UDP over IP for carrying voice and video streams. In circuit switched networks voice is sampled and encoded at 64 kbps μ- or A-law PCM (G.711) occupies 64 kbps of transmission bandwidth. IP carrying μ-law speech occupies 80 kbps of transmission bandwidth for 20 ms sampled packets (that's not including any link layer overhead), not to mention the overhead caused by sending RTCP packets. Clearly there isn't going to be a cost saving here!

RTP when carried over UDP is essentially unreliable, packets will get discarded for one reason or another, in this instance a compromise has to be made between lowering the gross bit rate and reducing the header overhead, by placing more samples in each packet, and increasing latency (waiting for the next 10 or 20 ms before sending the sample) and increasing the risk of loosing a sizeable chunk of the samples (remember one fragment lost, whole packet gone – poof!) and thus reducing listener quality.

5.7 IP APPLICATION SIGNALLING PROTOCOLS

This section very briefly describes the use of protocols that sit on top of IP for the control of multimedia communications. These protocols are the ones that control the connection of media streams and do for the packet-based world what ISUP, Q.931 and DPNSS, etc. do for the circuit switched world. There are four main call control protocols (if you will allow the term call control): the media gateway control protocols H.248/Megaco and Media Gateway Control Protocol (MGCP), and the peer-to-peer protocols SIP and H.323. Each of these protocols has a different heritage and history and we will explore each in turn in this section. To ease into the concept of packet-based control protocols I start this section with a look at a very simple control protocol RTSP.

Real-time Streaming Protocol (RTSP)

RTSP is described in RFC 2326 as an application level protocol for controlling real-time streams analogous to a TV remote control. Its primary use is for the remote control of media servers, for example a unified communications server that stores voicemail messages. The retrieval and playback of voicemail messages over an RTP stream can be controlled by RTSP.

RTSP has the following subset of messages: SETUP, PLAY, RECORD, PAUSE and TEARDOWN. The names of the messages are explanation enough of their intended purpose. Individual message streams are identified by an RTSP URL, not unlike the web URL, for example rtsp://mediaserver.acme.com:554/stream1.

The DESCRIBE message is used to request a description of the media stream. The response contains a description written in the session description protocol (SDP – more on this a little later). The converse message, ANNOUNCE, is used to post a description of a media stream to the media server and can be used in real time to announce a change to the description of a media stream.

RTSP's heritage in Hypertext Transfer Protocol (HTTP) is obvious if you examine the RFC, what is also apparent is its relationship with SIP (described later) because it too shares this common parent.

Media Gateway Control Protocols

In the Time Division Multiplex (TDM) network (Public Switched Telephone Network (PSTN) or Public Land Mobile Network (PLMN)) switches are organised around a switch plane, a Stored Program Controller (SPC) and trunk peripherals (see the explanation in the introduction to the section on circuit switching). To make use of IP connections to connect voice connections together, the decomposed circuit switch architecture has been created.

The reasons behind the decomposed architecture are many:

- the need to continue (for at least the short to medium term) to interface to the TDM network;
- continue to utilise all the existing millions of lines of code and man years of effort invested in the SPC software;
- expandability, the architecture allows a separation between the call control processing and voice trunking components, allowing each to be scaled separately;
- it opens up the opportunity for increased competition and vendor specialisations, standardised components and communications methods should allow for a multi-vendor environment;
- call control software can be moved on to the latest and greatest IT hardware and operating systems, boosting performance.

The decomposed circuit switch architecture removes the need for the switch plane function of the TDM switch and replaces this with an IP network. The original trunk peripherals that terminate the voice and signalling have become Media Gateways (MGs) and Signalling Gateways (SGs), respectively. The most important change has been the decoupling of the SPC to become the gateway controller.

In a circuit switch, the SPC is connected to an internal control bus that interfaces with the switch plane and trunk peripherals, since the circuit switch has been decomposed into component parts, the control bus must be extended to continue to control the peripherals. The control bus that controlled the peripherals has been replaced by media gateway control protocols. Figure 5.3 represents this logical decomposition. The point to note here is that whilst the diagram depicts the SGs and the MGs as separate components, in real-world implementations they may actually be the same physical box, it is the logical function that is separate.

The media gateway control protocols are essentially, what make this architectural decomposition possible. The two protocols MGCP and Megaco protocol have grown within the IETF and essentially Megaco is the culmination of the IETF work in the MEdia GAteway COntrol working group – MEGACO. MGCP was the earlier protocol proposal and incorporates work on Internet Protocol Device Control (IPDC) and Simple

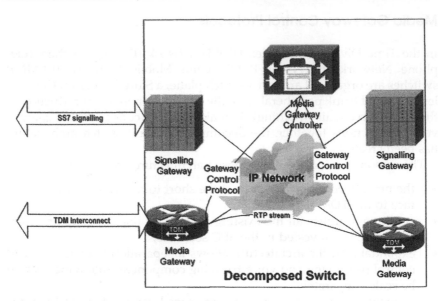

Figure 5.3 Decomposed switch

Gateway Control Protocol (SGCP). Megaco protocol on the other hand was the later proposal in the IETF and has followed parallel standardisation tracks through the International Telecommunications Union telecommunications (ITU-T) (possibly a first!) as H.248. The main common ground for both of them is that they are designed to be interior protocols, i.e. they only operate within the boundaries of the logical single voice over IP (VoIP) gateway, a number of other similarities exist (they would since work on the Megaco protocol integrated the work on MGCP).

The MGCP specification (RFC 2705) describes an Application Programming Interface (API) and a protocol for the control of the VoIP gateway by external gateway control elements. The standard refers to call agents that are the logical elements where the bulk of the call control function resides, essentially the gateway controller. It describes a master–slave protocol relationship between the call agents and the gateway functions, respectively.

RFC 2705 also gives examples of the gateways that might exist in the network, these are:

- trunking gateways, these are devices that interconnect to the existing PSTN or PLMN, Figure 5.3 shows these types of gateways;
- residential gateways, these are units that terminate Plain Old Telephony Service (POTS), BRI or xDSL lines and concentrate the traffic on to a backbone network, a good example of such a device would be a xDSL Integrated Access Device (IAD) or cable modem which has a

BRI interface, a POTS interface and an Ethernet connection for non-voice data;

- access gateways, these are components that reside on the Customer Premise (CPE) and interface to Private Branch Exchanges (PBXs);
- ATM gateway, in ATM terminology this device would be referred to as an Interworking Function (IWF) and converts ATM cells into packetised IP datagrams;
- network access servers, these are components that can be used to off-load dial-up modem traffic from the PSTN and terminal modem connections.

In MGCP media stream terminations are referred to as 'endpoints' and endpoints are linked together via 'connections', which can be point-to-point or multipoint. Connections are not distinct to any specific type of media and could equally be RTP or ATM AAL2 media streams or any other type of media stream or TDM connection for that matter. All the MGCP messages are text-based and transported over UDP.

It is the intention of MGCP networks to be able to provide the glue between PSTN telephony users and SIP or RTSP controlled sessions via the call agent. An example protocol exchange can be found in Section 2 (voice-based services), illustrating an MGCP exchange between a gateway and gateway controller, with a media server utilising SIP to connect an RTP voice stream between a gateway and itself.

The general opinion is that MGCP will be largely replaced by Megaco, MGCP has had implementation in networks, but most of the effort in decomposed gateways by the big players such as Nortel networks is focused on Megaco.

RFC 3015 is the common text version of Megaco protocol with H.248. The RFC incorporates and obsoletes RFCs 2886 and 2805; RFC 2805 being the original specification and 2886 changes and errata.

The Megaco specification breaks the components into a Media Gateway Control (MGC) layer and a Media Gateway (MG) layer. The MGC layer is where all the call control intelligence resides and is the master side of the control function. The MGC layer also has peer relationships between MGCs with SIP-telephony (SIP-T) being a proposal for this. The call agents referred to in MCGP reside in the MGC function. The MGC communicates with the outside TDM world via SGs. The SGs connect to the MGC via the SIGTRAN protocol SCTP (see later).

Megaco uses two main concepts to manage media streams: terminations and contexts. Terminations are the logical representation of the media streams to and from the packet network on the gateways, the protocol allows for signals to be applied to these connections and events to be received about the change in status of the media streams. Contexts are a means of group terminations (media streams) together to implement bridging and mixing of the media streams.

Sets of command primitives are defined that operate on terminations, to effective set-up and teardown media streams across the IP network between the media gateways. These are:

- Add
- Subtract
- Move
- Modify
- Notify
- Audit
- ServiceChange

Each of these commands can be grouped together into a single control message to a gateway. This grouping is referred to as a 'transaction'.

In order to accommodate a range of media gateways and to provide an extension capability Megaco defines 'packages' and profiles. A package is a group of properties, events, signals and statistics that a gateway can exhibit. A profile is essentially an application level agreement between the MG and MGC software over the use of a specific MG and communications between the MGC and MG. This is necessary to constrain the interface between a multitude of different gateways, because in a mixed MG environment, the MGC software must be able to control and understand each gateway's operating characteristics.

Phew! That was a very quick pass through gateway control protocols, if you need or want to know more, then I recommend you read the relevant RFCs and [WRIG].

SIP

Session Initiation Protocol (SIP) has taken both the Internet world and the telecoms world by storm; some have heralded SIP as the SS#7 of the future. Brushing the excitement aside, what is it all about? One word sums up SIP simplicity. SIP's ideas and ethos are based around simple protocol implementation. We've looked at the complexities of the SS#7 networks that have grown up over the past 30 years, and we've seen how the Internet can change things very rapidly. The Internet has achieved this by implementing simple ideas.

Back in the early 1990s a number of researchers were looking at implementing a network known as the MBONE, the multimedia backbone. To do this and utilise the multicast capabilities of IP they also needed some means of advertising and creating session between parties. Early work on the MBONE looked at announcing multicast sessions, which created the experimental standard – Session Announcement Protocol (SAP). In order to announce something they needed a useful way of representing the session. This was done through the Session Description Protocol (SDP).

An alternative to the 'announce to the world' approach is to explicitly invite each of the participants, to this end SIP was created.

The approach that SIP takes that is revolutionary from a telephony point of view is that SIP doesn't care where the participant that is being invited into a session is. This is unlike the PSTN, where you are explicitly trying to reach a person at a specific number (their telephone number) that has a relationship with a physical device. SIP uses Uniform Resource Locators (URLs) not unlike most people are now familiar with when they send email to someone, for example 'neillw@telecomsoapbox.org.uk'. Think about when you send an email, do you have any idea where the person is or for that matter care?

When you call someone you have to think, should I call them at home or the office or maybe I should just use their mobile number, wouldn't it be easier to just call one identifier all the time and let the network do the rest? This approach was tried as an Intelligent Network (IN) service without a great deal of success, I think as much because it was a premium service rather than it not being beneficial. To enable SIP to take this approach it makes extensive use of a component called a proxy server (Figure 5.4).

The proxy server's main role in life is to route requests for session participation (invitations) to the participants by utilising the translation from their SIP URL to their current physical location (for example an IP address). Clearly proxy servers need to be kept informed of the endpoint a particular URL relates to. This is done in SIP via a registration and update procedure and a registration server. More recently SIP application servers have also arrived on the scene. However, we're getting a little ahead of ourselves (see SIP CGI in Chapter 12).

SIP differs from the media gateway control protocols in that it is a peer-to-peer protocol and relies on intelligence in the end terminals. This approach is common in the Internet, where most of the intelligence is distributed at the edge. Where SIP differs from Megaco is also were it has similarities with H.323, even though they have different heritage,

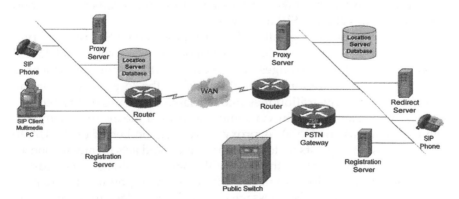

Figure 5.4 SIP architecture

and in fact numerous papers have been written comparing SIP and H.323.

SIP borrows a number of its elements from HTTP and SMTP. From HTTP it inherits URLs and client server connectivity. From both SMTP and HTTP, SIP gets its header style, and from SMTP text-encoded messages. The advantage attributed to text-encoded messages is ease of debugging, since they are readily understandable without specialist tools. SIP also borrows from multipurpose Internet mail extensions (MIME) for encoding additional non-text content to messages.

OK let's cut to the chase, how does SIP operate? SIP uses request methods (seven in all: REGISTER, INVITE, ACK, CANCEL, INFO, OPTIONS and BYE; with more cooking such as PRACK), and responses (six classes of response codes exist: Informational, Success, Redirection, Client error, Server failure and Global failure).

SIP as we have said is a peer-to-peer protocol, so whilst proxy servers are extensively used they are not completely necessary to set up communications. SIP messages are capable of being transported on UDP or TCP and in the case of UDP transport messages can make use of IP multicast for effectively creating an invitation to a multicast session (more than two participants). Just to get a feel for how SIP works Figure 5.5 shows a simple protocol exchange.

In SIP the application that supports the SIP end device is called the SIP User Agent (UA). The SIP UA is actually two components, a Client (UAC) and a Server (UAS). The UAC generates requests on behalf of the application and the UAC generates responses. In Figure 5.5 the UAC generates the INVITE and BYE messages, whilst the UAS generates the RINGING, OK and ACK messages. The SIP UAC is also responsible for encoding the session media parameters into a session description protocol payload as part of the session establishment.

A UA can be more than just a simple client and can be used to implement a gateway between different protocols. In this case the application is a piece of software that takes in messages from one protocol entity and uses the UA to generate SIP messages.

SIP servers come as logically SIP registration server, SIP proxy server and SIP redirect server. A SIP registration server accepts REGISTER messages; all other messages are rejected with a not implemented response. The registration server then makes the location information contained in the registration message available to the proxy and redirect servers in the same domain via some unspecified means (for example a common database or DNS). A proxy server acts on behalf of the UA to forward session requests on to their destination and acts on any responses by returning them back to the UA (with some exceptions). The redirect server essentially performs a not here function by responding to a request with a try somewhere else message containing the place to try, it is then the responsibility of the UA to regenerate the request message and send it

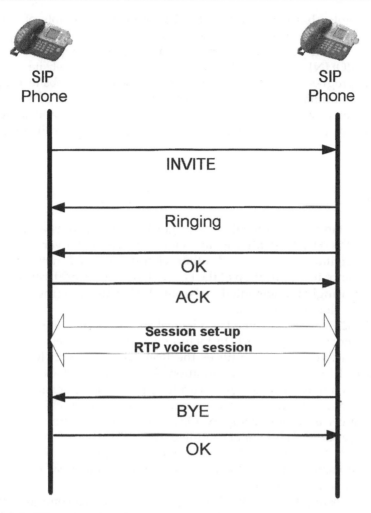

Figure 5.5 SIP protocol exchange

to the new address. These logical server functions can be separate, or can (as is common) be grouped into a single physical entity.

Proxy servers can also implement services on the behalf of the users they look after. For example one proposal has the REGISTER messaging being able to upload a script. If a combined registration server and proxy server function also implemented a scripting engine that could execute these scripts, then rules could be applied in the network to support user-based session profiling. This type of approach is how SIP application servers operate.

The SIP specifications can be extended to support extra methods for example an Internet draft from SIP extensions to support presence service (see Chapter 12) proposed to new methods: NOTIFY and SUBSCRIBE.

This chapter has covered only the basics on SIP and lots of sources exist that cover both the existing standards and the work in progress. I strongly suggest you use the explanations here as a stepping-stone and look up [JOHN] and the SIP standards on the IETF website and the work in progress reflected in Henning Schulzrinne's website (http://www.cs.columbia.edu/sip/). SIP looks to be the future of signalling protocol for advanced next-generation network services and a significant amount of effort is being put in around the world on its development.

H.323

The H.323 family of protocols was originally developed as an enhancement of the H.320 standards for videoconferencing over integrated services digital networks and other circuit switched networks and services. Since the ratification of the H.320 standards in 1990, corporations have increasingly implemented LANs and LAN gateways (routers) to the Wide Area Network (WAN).

The H.323 standard evolved as a logical and necessary extension of the H.320 standardisation effort, to include corporate intranets and packet switched networks. H.323 utilises the Real Time Protocol (RTP/RTCP) from the IETF, along with internationally standardised CODECs (see Chapter 2 on voice digitisation). With the ratification of version two and ongoing standardisation effort on version three and four, H.323 is slowly getting larger and more complex. One of the issues version three deals with is inter-domain (or zone) routing of calls. The early releases of H.323 assumed all the communications endpoints belonged to a single administrative domain; however, the use of H.323 in a wide area environment meant this had to be fixed.

H.323 is being extensively used for voice and video communications over the Internet the most common implementation being that of Microsoft's NetMeeting™ product. H.323 is an umbrella protocol specification, in that it covers a collection of protocols, the relationship between the different specifications that make up H.323 is shown in Table 5.1.

All of these protocols work in concert in H.323-based systems to carry out all the operations necessary to perform a multimedia conference. Whilst all these specifications exist it is not necessary for a H.323 system to support all of them, in fact many implementations only support a subset.

Figure 5.6 shows the relationship of the protocols as a protocol stack. The first thing to note is the absence of layer two (link layer) and layer one (physical layer). This is because H.323 doesn't have any preference over what these should be. The other point is that the diagram (artificially) separates out the control/signalling protocols from the media; this is

Table 5.1 H.323 standards

Feature	Protocol
Call signalling/connection management	H.225 (Q.931 messages)
Media control	H.245
Audio codecs	G.711, G.722, G.723, G.728, G.729
Video codecs	H.261, H.263
Data sharing (shared whiteboard)	T.120
Media transport	RTP/RTCP over UDP on IP
Conference control/management	H.332
Security (RAS)	H.235
Circuit switched interworking	H.246

purely to show more clearly the distinction.

Figure 5.7 shows the relationship between the logical components that make up the architecture of a system/network based on H.323.

These elements are terminals, gateways, gatekeepers and Multipoint Control Units (MCUs).

Terminals are often referred to as endpoints, they provide point-to-point and multipoint conferencing for audio and, optionally, video and data.

Gateways interconnect to the PSTN (could be mobile as well as fixed) networks for H.323 endpoint interworking or other different networks for example a SIP-based network.

Gatekeepers provide admission control (security and validation) and address translation services for terminals or gateways. In version three of the specifications gatekeepers also communicate with each other to manage connection of sessions between zones. Figure 5.7 shows two gate-

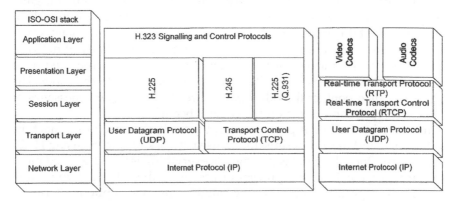

Figure 5.6 H.323 protocol stack

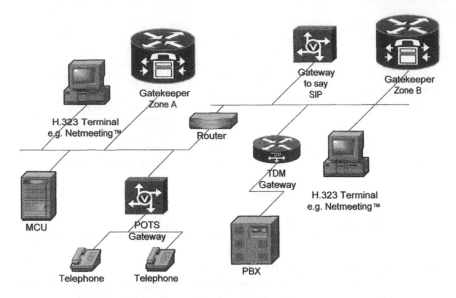

Figure 5.7 H.323 components

keepers each managing a separate zone (A and B). The figure depicts each zone being separated by a WAN connected via a router. This isn't necessarily what determines a zone boundary. In fact zones can span wide areas and are purely for administration purposes.

MCUs are devices that allow three or more terminals and gateways to conference with either audio and/or video sessions. The MCU is made up of a Control unit (MC) and an (optional) Multipoint Processor (MP). Essentially the control portion ensures all connected devices have negotiated compatible capabilities and the MP is the mixer that mixes all the different streams together.

H.323 call establishment is a fairly involved process (see Figure 5.8), which starts with the validation of resources and permissions by the gatekeeper. This process is triggered by a request (using H.225) from a client wanting to establish a connection to another party. After the gatekeeper has validated the credentials of the caller and the availability of network resources, the IP address of the far end (called party) is returned (or a gateway if the party is on a different type of network) to the caller. The caller can then contact the other party directly using Q.931 messages or the gatekeeper can do this on the caller's behalf.

In order for the called party to accept the call, they need to contact the gatekeeper and ask for permission to use network resources (using H.225 messages). Once the gatekeeper signals everything is OK, the called party can accept the call. In the process of accepting the call, the two parties' terminals negotiate voice coding algorithms and options (terminal capabilities). Once the negotiation reaches a satisfactory conclusion the

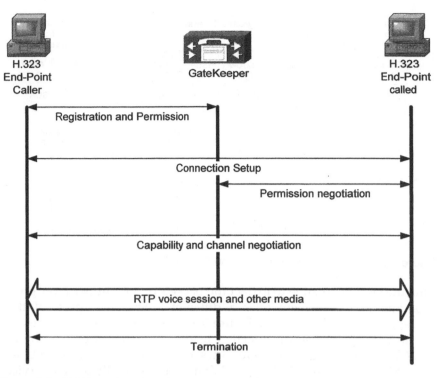

Figure 5.8 H.323 protocol exchange

two parties can communicate.

This simple description hides a lot of complexity present in the H.323 protocol suite, and the reader wishing to understand fully this complexity is referred to the appropriate standards documentation.

Stream Control Transmission Protocol

SIGnalling TRANsport (SIGTRAN) is a working group within the IETF, whose role is to define standards that allow the transport of SS#7 messages across an IP network. Their main work, defined in RFC 2719 (Framework Architecture for Signalling Transport) covers the need for a reliable transport mechanism (called the signalling transport in the standards document) to carry SS#7 messages across an IP infrastructure. I recommend you re-read Chapter 1 on SS#7 signalling before proceeding.

The main specification is that of the Stream Control Transmission Protocol (SCTP), which defines a protocol for the reliable transport of protocols such as Q.931, ISUP, TCAP and MAP. Remember from the section on SS#7, SS#7 is actually a packet-based protocol that is carried (as a matter of convenience) in 64 kbps bearers in the current circuit switched networks.

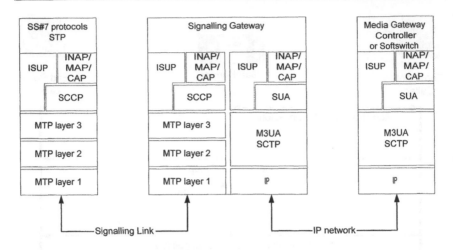

Figure 5.9   Signalling gateway function for SCCP

This means it is possible to extract the packets of SS#7 upper-layer protocols and transport them over a different lower layer protocol than
Message Transfer Part (MTP). Figure 5.9 shows the conversion necessary
for an SS#7 to IP gateway function using the protocols defined by the
SIGTRAN group. The MTP layer 3 User Adaptation Layer (M3UA) is a
layer for carrying any SS#7 protocol that would normally use the services
of MTP layer 3. This in essence is SCCP and ISUP (plus international
variants). M3UA forms a layer between SCTP and the SS#7 application
part protocols.

The SCCP User Adaptation layer (SUA – Figure 5.9) is used to transfer
any SCCP user part protocols (INAP, MAP, CAP, etc.) over SCTP.

The work of the SIGTRAN group is ongoing at the time of writing and
the protocol specifications they produce are extremely important to the
integration of next-generation networks with the current PSTN (fixed and
mobile). As can be seen in Figure 5.9, the SCTP is used to transport
signalling from the decomposed signalling gateway (SG) to the gateway
controllers defined in MEGACO and MGCP and the more generic softswitch (see Chapter 10). It can also be used to transport INAP messages
from an 'IP-enabled' SCP (Section 10.5 on voice services has an example of
this).

Work is also still ongoing to define a set of adaptation layers that are
used to transport the signalling information using the services of SCTP
(examples of which are M3UA and SUA, above). The adaptation layers
also provide valuable services such as protection against masquerading
(other devices pretending to be trusted devices), fraud and non-repudia-
tion, congestion control and other generic interfaces for non-signalling
applications.

As these standards are still under development I suggest the reader consults the most up-to-date information on the SIGTRAN working group page at the IETF website (http://www.ietf.org/html.charters/ sigtran-charter.html).

Multi Protocol Label Switching

This section is about Multi Protocol Label Switching (MPLS), it is also a means of discussing some of the salient points about being able to utilise MPLS for quality of service in IP networks, and traffic management. We discussed in the chapters on circuit switching that voice quality is maintained both by having a clear channel connection between the two points in the network and by providing synchronisation of the voice streams in the TDM bearers. This approach is great for circuit switched networks, but of no use in a packet switched environment where as we have discovered in the section on real-time transport protocol (Section 5.6) no connection actually exists between the endpoints and the route a datagram can take in conventional routing can be different from packet to packet not to mention that timing/synchronisation is not possible with buffers in routers all over the place making a mockery of any kind of consistent delay. MPLS is one of the mechanisms we can use to address this problem (more a little later).

We will also discuss in the chapter on ATM (see Chapter 7 on voice and data convergence) how apparent timing can be constructed. We also discuss how virtual connections can be created with ATM in the form of Switch Virtual Circuits (SVCs) and Permanent Virtual Circuits (PVCs). MPLS has a lot in common with ATM in respect to virtual connection and we will highlight some of the benefits of the combination of ATM and MPLS and how they are similar.

MPLS has had a fairly interesting history with different manufacturers and different parties in the IETF looking at how to more closely integrate ATM and IP. Various issues exist when you connect routers together in a meshed network, the main one being maintaining the link status and routing tables, when a number of routers are connected in a fully meshed configuration. This type of configuration is common when using a collection of VCs from an ATM cloud. The configuration can cause an explosion in the number of routing update messages should anything in the topology change, such as a Virtual Connection (VC) failing, just think of the number of neighbours a router has (actually $n-1$, i.e. the number of routers in the mesh minus yourself). You can try to cure this problem by reducing the meshing and using an intermediate router to forward packets to routers that are not directly connected to a specific node, however, this intermediate router can quickly become a bottleneck if the volume of traffic increases.

The alternate approach proposed was to add routing capability to the

ATM switches, thus making them part of the network and mapping the IP routes on to either the Virtual Path Identifier (VPI) or Virtual Channel Identifier (VCI) parameter in the ATM cell header, thus creating a Label Switch Router (LSR). This means the routers at the edge of the network now only have one adjacent node, the ATM switch, thus reducing the scaling issue. This is a very simplistic view, but I hope it explains the idea. By making this leap of faith and allowing the ATM switch to participate as essentially a router, but with added capability in the form of being able to aggregate routes on to specific virtual channels, it creates the interesting property of being able to choose much more carefully the 'route' the IP datagrams follow.

The commonly used diagram that represents the ideas of label routing is the 'fish' diagram, because it resembles a fish. Figure 5.10 shows the 'fish' diagram and router C is a label switch router. Datagrams from router A destined for router F could in ordinary routing take either of the paths from C to F (i.e. via router D or router E). The same is true of datagrams from router B. However, because router C is a label switched router, the decision can be made to route all datagrams from router A via E and all diagrams from B via D.

The aggregation of traffic on a specific route creates one of the key properties of MPLS that of traffic grooming and engineering. By choosing to map tuples of elements from the IP header (source address, destination address, ToS field even protocol type) onto a specific 'label', traffic of different types (ToS), from different applications (protocol type), etc. can be mapped to different routes, effectively creating a forced route through the network for specific criteria. Another property that can be created, for example is a routing decision is made to map specific labels on to specific output queues or to assign priority to specific labels and insert datagrams higher up (or lower down) a specific queue. Thus if this latter approach is used to map IP quality of service capabilities (int-serv and diff-serv – see the work of the IETF integrated services over specific link layers and differentiated services for details of work on IP QoS) on to specific labels, MPLS can be used to support IP QoS networks.

The speed of transfer of datagrams can also be improved if all the label

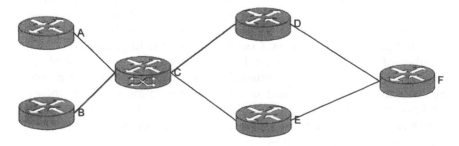

Figure 5.10   MPLS fish diagram

switch router is doing is looking up a memory area based on a simple label to determine the outbound port the traffic should egress on. Clearly this implies the assignment of the label has already been performed by some control mechanism, for example pre-provision like PVCs or allocated based on some form of resource reservation (for example using int-serv and RSVP).

OK so that's some of the reasons for considering a label switching approach, why is MPLS called multi protocol and how does it work? Let's start with a 'label'. A label is a fixed length field with no specific structure, it doesn't encode say the IP address of the destination or any other combination of IP header fields or for that matter any other protocol fields (IPX, NetBEUI, etc.). It just contains a label value. In essence a label switch router just keeps a look-up table indexed by the value of the label extracted from the incoming datagram.

MPLS is multi protocol, because the label is embedded between the link layer header and the network layer header on network layer protocols that don't natively support a label (Ethernet for example) and in the case of protocols like ATM the label is placed in the cell header in either the VCI or VPI fields depending on the implementation. Since we have already stated the label has no specific semantics in relation to the network layer protocol, then clearly MPLS can support any protocol.

MPLS can handle unicast and multicast quite simply by the structure of the forwarding table in the router. For example the table held in memory could be a series of linked lists indexed by the label extracted from the incoming datagram, label 'n' corresponding to row 'n' in the table. With each entry having attributes such as the address of the next hop, the outgoing label value, queue (buffer) to place the datagram on an outbound port (see Figure 5.11). Literally all the label switching function does is extract the label from the incoming datagram, use it to look up an entry in the table, select the row and traverse the list extracting the values (attributes if you like) and placing the datagram in the queue on the outbound port with the appropriate label.

Hopefully this explains a little about MPLS and its capabilities. One important thing MPLS is commonly linked to is virtual private networks (VPNs). Hopefully from the explanations above it is simple to see how a VPN can be constructed by the pre-provisioning of specific labels to different customers, essentially creating a collection of segregated flows through a label switched network.

Hopefully you can see how QoS can be supported by assigning labels to specific quality parameters such as assigning values from the Type of Service (ToS) field in an IP packet header to specific labels and then using this label to index a table that places the traffic of different types of service in different output queues or higher up a single output queue. You can also see how if an ATM network has been constructed with

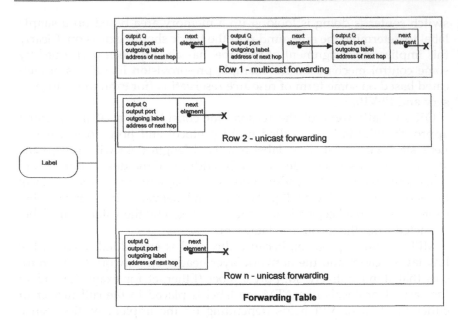

Figure 5.11   Label switch router example forwarding table

specific ATM Adaptation Layer (AAL) characteristics how very simple it is to assign MPLS labels to ATM virtual connections and thus map, say IP QoS to ATM flows.

In closing MPLS is a very simple but powerful technique for speeding up and controlling the routing of traffic in a network. If you would like to know more, then I recommend you read [DAVI].

6

Access Technologies

6.1 INTRODUCTION

Access is arguably the most important part of the network infrastructure, since it is the point at which all revenue-generating customers come into contact with the network. It is also an area of great diversity from basic telephone access through private exchange and Automatic Call Distributor (ACD) connections, not to mention HFC cable Integrated Access Device (IAD) protocol stack TV, to leased data service connections and wireless connectivity in the form of wireless Local Area Networks (LANs) and of course the now near ubiquitous mobile phone (in Europe at least).

The one major advancement that is without doubt the greatest improvement in both functionality for the customer and revenue generation from the network operators' perspective is the introduction of xDSL technologies that have increased the capacity of the humble copper pair of cables in the local loop. This has also created major ramifications across the world with the 'unbundling' of the local loop via regulatory control causing severe pain to the incumbent telecoms operators.

This chapter covers the area that has undergone as much change in its history as the change to the core networks that this book discusses. The invention of the telephone by Bell (as discussed in the Preface) started the access infrastructure from the four and two-wire circuit carrying analogue voice, to the capabilities of xDSL. Co-axial cables and fibre optics have seen an increase in the data carrying capacity of the local loop from a few hundred bits per second to the thousands of megabits per second capability of fibre optic cables.

The increase in capacity is what is allowing the increase in the complexity of services that can be offered from the core of the network and has allowed the proliferation of the Internet model of intelligent endpoints.

Until now the opportunity for enhanced services has been encumbered by the low capabilities and features of the humble POTS handset. Plain Old Telephony Services (POTS) is the name give to the basic service provided by an analogue handset and the service capabilities of the Dual Tone Multifrequency (DTMF) signalling in the form of services such as call waiting and three-party transfer.

Pretty Amazing New Services (PANS) are all the services that are enabled by the new technologies such as voice recognition and convergence, basically all the services I will cover in Section 6.2. The touch point for all these services is the access network and this chapter covers the important aspects of how this is being enhanced.

6.2 INTEGRATED SERVICES DIGITAL ACCESS

Arguably the first change that enabled greater capabilities in the form of signalling (loop disconnect and DTMF being the POTS signalling) was the move to a digital access mechanism on the copper pair into the home and businesses. This digital access mechanism was basic rate integrated services digital network access. The business equivalent was primary rate Integrated Services Digital Network (ISDN). ISDN is built on top of the signalling system number 7 (SS#7) network described earlier; it is this foundation that created the foundation for the convergence of voice and data services. The heritage of SS#7 means that ISDN is based on connection-oriented communications, utilising the Q.931 signalling protocol for basic call control and the Q.932 and Q.950 specification for supplementary service procedures and protocols. Additionally in order to allow for *'feature transparency'*, the carriage and correct operation of features invoked in the access network through the integrated digital network, the facilities provided by the ISDN signalling must be mapped to the ISDN User Part (ISUP) signalling.

Basic rate ISDN for Small to Medium Enterprise (SME) and home use is now prevalent as a narrowband service. It allows access to two 64 kbps *bearers* or B-channels and a 16 kbps signalling channel. The B-channels can be used to carry switched speech in digital encoded form (see the section on speech encoding), or can be bonded together to form a 128 kbps switched data link using multi-link Point-to-Point Protocol (PPP). The signalling channel could also be used for 16 kbps packet switched services in the form of X.25, however, this capability has never been fully utilised.

The primary rate is delivered as E1 or fractional E1 service. This is in the form of 384 kbps (H0, 6 × 64 kbps bearers), 1536 kbps (H11, 24 × 64 kbps, delivered as T1) and 1920 kbps (H12, 30 × 64 kbps, delivered as E1). Access to the basic rate capacities is through the S and or T access points, additional devices in the form of network terminating equipment (TE) is used to connect to these interfaces. For example a terminal adaptor (TA) is used to connect basic analogue handsets. The S and T interfaces actually allow for the connection of up to eight TEs.

When the specifications for ISDN and the Integrated Digital Network (IDN) were being put together, 128 kbps was thought sufficient to carry advanced services such as slow scan video. The design principles behind the IDN were no different from the desires of the current raft of new services, what has changed is the bandwidth requirements and the mass acceptance of packet-based systems over circuit switched systems. The initial specifications for ISDN are now referred to as Narrowband ISDN (N-ISDN) and a new set of specifications were drafted around Asynchronous Transfer Mode (ATM) that were called broadband ISDN. These broadband specifications have largely been augmented in the access area by Digital Subscriber Line (DSL) technologies that incorporate ATM at their core, more on this a little later.

The uptake of ISDN was initially slow as the requirements for high-speed access at the time and the price of switched connections made the service prohibitive. How things have changed, the Internet pricing has enabled free calls for basic rate ISDN connections, allowing a semi-permanent 64 kbps connection to the Internet to be created, effectively creating a leased 64 kbps circuit between a home user and their Internet service provider. There was a time when leasing a 64 kbps connection could only be justified by companies, now for a few pence per minute a 128 kbps connection to the Internet for the home user is possible, and as DSL takes off even faster access will be possible.

ISDN is made possible on the copper pair by the use of echo cancellation techniques that allow bi-directional digital data to be transmitted directly over the copper pair. The main downside to ISDN is it is still heavily reliant on a circuit switched network infrastructure to carry the bearer channels, that's because of its heritage in the circuit switch world. When these bearers are combined and PPP is used to encapsulate data, two or more circuit switched connections are required across the IDN to carry the data traffic to a packet network interconnect point such as a router, this is expensive of resources in the circuits switched world. It is this last point that will largely cause the demise of ISDN as an access mechanism, and the uptake of xDSL technologies (regulation, politics and cost aside) as the replacement for copper loop access.

Basic rate ISDN presents itself in the UK as a little grey box with a green light fastened to the wall. The S-Bus connections are normally presented as two eight-pin connectors. These connections can use standard category 5 cabling to connect equipment up to 200 m away.

This very brief chapter has covered only superficially the aspects of ISDN as an access technology for convergent services. ISDN will be with us for as long as the circuit switched network is present, and no doubt will remain as the one technology that will extend the life of the Time Division Multiplex (TDM) based network as a revenue generating cash cow. One closing remark on ISDN, Chapter 5 on packet switching mentions frame relay in its introduction, it is interesting to note that the original work on frame relay specification was as a means of carrying packet services from an IDN. The basic premise of frame relay in this context was a simplification of the requirements and overhead imposed by the X.25 specifications. Frame relay has taken on a life all of its own as a narrowband data service. It is interesting to note that the work on broadband ISDN and in particular its link to the ATM forum has given the industry another means of carrying high-speed data and voice communications in an integrated fashion. If you would like a more comprehensive study of ISDN and its relationship to frame relay and ATM I suggest you try [STALL].

6.3 DIGITAL SUBSCRIBER LINE

Why did I follow a description of ISDN with that of DSL? One answer may have been because it is the next step in capacity for the access network, good answer but not the original motivation. DSL shares a relationship with ISDN – ATM – and as such is the logical delivery mechanism for the ideas behind broadband ISDN. In fact one of the original motivations for DSL was Video on Demand (VoD) services, as a response from the copper-line-based telcos to companies like Time Warner trialing VoD over cable TV networks, this gave birth to the Asymmetric DSL (ADSL) specifications. Interestingly the demand for VoD dried up (a trip to the video store being cheaper) and it was the growth in demand for Internet access that breathed new life into the digital subscriber line.

The author believes DSL will replace ISDN as the data and voice access mechanism for both consumers and small businesses. The cost model for always on connections will continue to decline through the use of this technology and through regulatory forced competition in the local loop. IDC have pretty bullish forecasts for broadband growth over the next 5 years. Whether IDC's forecasts will be met have as much to do with the strength of the regulators vs. the incumbent telecoms operators as they do the viability of the technology. The growth in the DSL broadband market has an, unfortunate some might say, link with the Unbundling of the Local Loop (ULL) and it is the fracas over unbundling that has and is causing severe pain for both incumbents and competitive local carriers.

Digital subscriber line is actually a family of physical layer specifications (referred to as xDSL) defining date rates from around 300 kbps up to around 50 Mbps. This family of technologies are Asymmetric DSL (ADSL), Symmetric DSL (SDSL), High bit rate DSL (HDSL), Very high bit rate DSL (VDSL) and just to add to the alphabet soup, there is even an ADSL-lite.

DSL utilises the copper pair to residences in the same way that ISDN utilises the copper pair. Both of the technologies take advantage of digital signalling processing and modulation techniques to increase the data rate available from the copper pair. Transmission rates of all these technologies are limited by the quality and distances involved in delivering the copper pair to the premises. So whilst high rates are theoretically possible, in practice real-world rates are much lower.

ADSL, as the name suggests, is faster in one direction than the other. Data rates from the network to the premise equipment can be up to around 8 Mbps, whilst the data rates from the premise equipment up to the network are around 600 kbps (line conditions permitting). In the real-world implementations, rates of around 1.5 Mbps in the down link and 128 kbps in the up-link are possible. ADSL provides the ability to vary the data rate according to line conditions. At the time of writing ADSL is the most popular choice for telcos to install as a residential and SME broadband service, competing with the cable TV companies who are rolling out cable modems.

SDSL provides the same data rates in both directions and is seen by many as a replacement for leased line services. In practice data rates of the order of 2 Mbps are possible, but unlike ADSL the data rate is fixed and requires (generally) better line conditions. At the time of writing a number of new entrant telecoms network operators, ITSPs and Internet Service Providers (ISPs) are looking to SDSL to provide SMEs with both voice and data services.

The backhaul connections to DSL lines are provided in most cases in the form of a multiplexing switch combined with a DSL modem – DSL Access Multiplexor (DSLAM) in the local exchange or point of presence. This multiplexor is used to concentrate the endpoint traffic down on to a backbone transmission network using statistical multiplexing. It is the concentration factor of the DSLAM (10:1, 20:1 and even 50:1) that can cause the most problems to real-time applications. This concentration is also known as contention and is used to reduce the cost of provision of large backbone pipes. The launch of voice services on SDSL will mean new and existing carriers will need to consider carefully the design of their backhaul networks.

The focus on the SME and consumer market for xDSL technologies means the device in the customer premise, normally referred to as an Integrated Access Device (IAD) or in these circumstances it is also referred to as a residential gateway, needs to be simple to set up and operate. The IAD for consumer and SME premises is responsible for providing a more *user-friendly* connection to the broadband infrastructure. The IADs in use at the time of writing present either an Ethernet connection or a Universal Serial Bus (USB) connection. The IAD is actually a combination of an xDSL modem and other components for example: an ATM protocol stack, an Internet Protocol (IP) stack (including PPP and possibly Point-to-Point Tunnelling Protocol (PPTP) or L2TP) and IP router/firewall. Figure 6.1

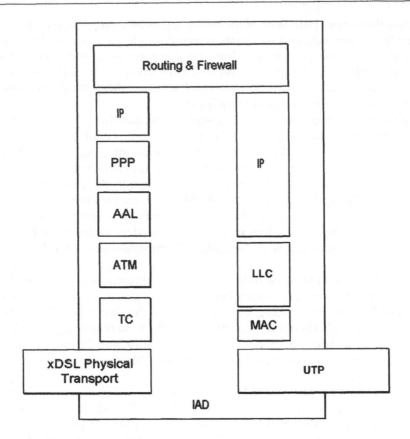

Figure 6.1 Typical IAD protocol stack

depicts the typical protocol stack for such a device (see the section on voice and data convergence for a description of the ATM layers – TC, ATM, ATM Adaptation Layer (AAL)). The drawback of this type of gateway is that the ATM layer and quality of service (QoS) capabilities are hidden behind the gateway. This could be remedied by for example adding Multi Protocol Label Switching (MPLS) to the gateway's capabilities and using specific 'differentiated service' types to map particular quality requirements on to MPLS tags and then mapping these on to ATM virtual circuits.

An important factor in the delivery of DSL to residential customers with only a single copper pair is the ability to ensure a POTS handset will work even during power failure of the IAD. This is to ensure calls to emergency services can still be made if the DSL equipment fails. This capability is normally provided via a POTS splitter, but battery backup on the IAD is also a possibility (Figure 6.2).

The original intent of xDSL as a broadband data access service has resulted in debate over the carriage of voice over xDSL (beyond the ability

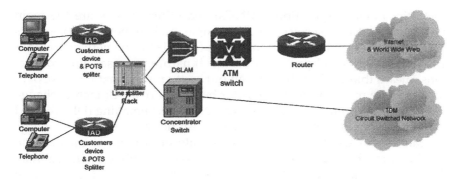

Figure 6.2   Generic xDSL architecture

to use POTS splitters, which aren't possible with all the xDSL technologies). Some argue for Voice over DSL (VoDSL) using ATM because of its in-built QoS capabilities, whilst the other camp debate for Voice over IP (VoIP) directly from the IAD. The other factor that can greatly affect voice quality is the contention for the backbone network from the DSLAM. Network operators looking to deliver voice from their xDSL networks or partner networks need to consider carefully the service, bandwidth and latency issues caused by concentrating the traffic.

The ATM forum has produced a specification called loop emulation over AAL2 for the carriage of voice and multimedia over ATM [LES]. The specification describes a mechanism for transporting voice (compressed or otherwise with silence removal), ISDN B and D channels and fax over xDSL (and cable or wireless links). It effectively also provides for (depending on the implementation of the IAD) the backward compatibility of xDSL to BRI. This and other specifications are being worked on to deliver voice over xDSL.

The transport of voice services over DSL (at the time of writing) still requires work before services will be offered, however, the author believes there is sufficient market opportunity for ITSP and new telecom entrants (as well as existing and incumbent operators) to want to offer the service. The opportunities for service and the service architecture will be explored in the next section. If you would like to learn more about xDSL then look at [WARR] as a good start.

6.4   LEASED LINES AND OTHER FIXED LINE SERVICES

One of the existing revenue streams for telcos is leased line services. Customers can lease clear channel services from 64 kbps upwards to 155 Mbps. The lines are a private dedicated connection between two points across national and in some cases international boundaries.

Multinational Corporations (MNCs) and other carriers are the main customers of these services. For example mobile network operators often lease large amounts of bandwidth from fixed line network suppliers. The clear channel private (dedicated) nature of leased line connections, means that the leaseholder can put pretty much what they like over the links. The most common use is for building Private Branch Exchange (PBX) networks utilising Digital Private Network Signalling (DPNSS) and Q.SIG signalling to construct a private transit network of switches. Large companies can use these networks to route both internal calls and to provide a mechanism for toll-bypass, by routing Public Switched Telephone Network (PSTN) destined calls to 'break-out' of the private network in the local area of the destination of the call, thus avoiding long distance call charges.

MNCs also use these services to construct private data networks and will use E1 and E3 circuits to link data centres together.

One of the issues surrounding the delivery of xDSL technologies is the lower cost of xDSL compared to leased line services, telcos are being very cautious over the price sensitivity of releasing DSL too soon and causing themselves a loss of revenue from their leased line services.

7

Voice and Data Convergence

7.1 INTRODUCTION

The subject of this book is all about the convergence of voice and data services into a new world of advanced (and hopefully useful) applications that will enrich the way people work and communicate. So a specific chapter entitled voice and data convergence could jar a little with the reader, however, this chapter was prompted by the words in the International Telecommunications Union telecommunications (ITU-T) I.121 specification "The need to integrate both circuit- and packet-transfer mode into one universal broadband network". It is interesting to reflect that what we now start to see becoming achievable through Internet Protocol (IP) was stated as a goal in an ITU-T specification in the late 1980s.

One of the difficulties in finding a place for Asynchronous Transfer Mode (ATM) in the first part of the book was that ATM is a packet technology, but has a circuit switching heritage, so if the reader will forgive me, I have chosen to pick up ATM under a chapter all of its own under voice and data convergence.

The simple statement (amongst others) in I.121, gave rise to the specifications for Broadband ISDN (B-ISDN) and through the work of the ATM forum the specification of asynchronous transfer mode. This chapter explores ATM as a key enabler to convergent communications and applications. ATM has suffered against transport control protocol/Internet protocol (TCP/IP), as TCP/IP started getting all the attention and ATM seems to have been backwatered as a transmission mechanism.

That's not so say ATM isn't a success, quite the contrary, ATM has had a resounding success and in fact has found its way into public data networks in a big way and may continue to gain success as a mechanism for creating connection-oriented paths through a network that carries Multi Protocol Label Switching (MPLS) (tag-) routed IP packets. MPLS and ATM share a common goal, the idea of being able to relay packets at high speed with quality of service. Some might argue that this is a waste and in fact, in the introduction on packet technologies I state that IP, combined with MPLS routers, could be carried directly over Dense Wave Division Multiplexing (DWDM). The truth of it is that ATM, MPLS and IP will be intimately bound together over the next few years and maybe as technologies for creating even higher speed MPLS routers emerge, then ATM will get left behind?

7.2 ASYNCHRONOUS TRANSFER MODE

Asynchronous transfer mode is a packet switched (or more accurately cell switched) technology that has evolved from a circuit switched world! This inheritance means ATM is a connection-oriented technology designed for use in the backbone of carrier networks. Its Integrated Services Digital Network (ISDN) and Signalling System number 7 (SS#7) heritage means that the standards define two different types of protocol for connection set-up and teardown. An interior protocol called broadband ISDN user part (B-ISUP, cousin to the SS#7 call connection protocol ISUP) for the connection between ATM switches (called the Network Node Interface (NNI)) and an exterior protocol Q.2931 for the outside edge of the network (called the User to Network Interface (UNI)). The standards also go much further than this and go to great length to build in manageability, again based on its heritage without the management aspect ATM wouldn't have got anywhere in a telco environment and of course quality of service capabilities.

We start our look at ATM with the B-ISDN reference model, because this is where it all started. The B-ISDN reference model is based on the Open Systems Interconnection (OSI) reference model (remember we mentioned this in the section on SS#7, Chapter 1. It's incredible where it just keeps turning up) and the original work on the ISDN standards. The B-ISDN model contains three planes (columns if you like): the User plane (U-plane), the Control plane (C-plane) and the Management plane (M-plane) (Figure 7.1).

The C-plane is where connection control resides; it is here that you will find Q.2931 and B-ISDN for control of connection set-up and teardown. The B-ISDN model supports both a Switch Virtual Circuit (SVC) and a Permanent Virtual Circuit (PVC) connection mode. The C-plane is responsible for the management of the SVCs. PVCs are constructed via maintenance action and are thus the responsibility of the M-plane.

The M-plane is responsible for operations, administration and mainte-nance. The M-plane also has responsibility across the other planes and between the planes to ensure management function can monitor and control all the aspects of the B-ISDN *stack*.

The U-plane is simply the place where application level protocols and functions reside. In this context TCP/IP is counted as an application level protocol to the B-ISDN service. Other examples of elements/services that reside in this plane are video on demand and Voice over IP (VoIP).

Sitting below each of these plane functions are two additional layers, the ATM layer and the physical layer. Between the ATM layer and the U-, M- and C-planes is a glue function called an adaptation layer that allows these planes to utilise the cell-based structure of ATM.

The Signalling ATM Adaptation Layer (SAAL) glues the C-plane to the ATM layer. The ATM Adaptation Layer (AAL) functions for the C-plane are further segregated into AALs based on the type of service/traffic the connection (SVC or PVC) will carry (this will be expanded on later). The M-plane also uses an adaptation layer to carry management information and control protocols between the ATM nodes in the network. Examples of these protocols are the Simple Network Management Protocol (SNMP),[1] Common Management Information Protocol (CMIP)[2] and Local Management Interface (LMI).[3]

There is a subtlety in the use of the AAL, for the U-plane, it only operates at the edge of the network not in the core, for the C- and M-planes the AAL is used both at the edge and in the core. For example SAAL is used to carry UNI signalling (Q.2931) at the edge and to carry NNI signalling (B-ISUP) in the core.

How are different media with different loss, delay and bandwidth requirements carried across a fixed cell size network (we'll come to the fixed cell size later)? This is the role of the adaptation layer. As we have already seen voice has a number of constraints with regards to delay and jitter, the same is true for high quality video-on-demand services (some of the video-on-demand requirements can be overcome with other techni-ques). Data services such as email and web browsing are less demanding of the network, however, loss of data for these services is a problem, whereas loss of the odd voice or video sample is less of an issue. The AAL has to deal with these differences and be able to place all of these different requirements into the ATM fixed 53-byte cell structure.

The ATM standards organisation decided that the ATM adaptation layer be split into two sublayers: the Convergence Sublayer (CS) and

---

[1] An Internet standard for exchanging management information between network nodes and a network manager.
[2] ITU-T standard for exchanging management information between network nodes and a network manager.
[3] Used in frame relay networks for exchanging status information between devices such as routers.

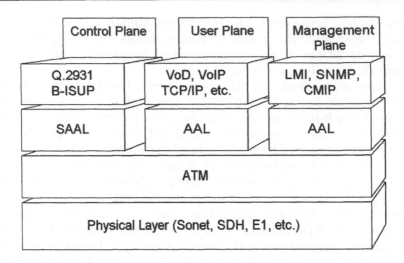

Figure 7.1 B-ISDN model

the Segmentation And Reassembly sublayer (SAR). As the name suggests the SAR sublayer is responsible for breaking the application level payload down into 48-byte chunks to fit in the cell. The application level data may be anything from 1 byte to many thousands of bytes (remember an IP packet can theoretically be in excess of 65,000 bytes in size). The ATM payload may not carry all application level information there may also be some adaptation layer headers and trailers. A series of classes of traffic have been defined by the standards (classes A–D and X), each class represents the characteristics of the application traffic it represents: whether it is synchronous or asynchronous, whether it is a constant bit rate source (voice) or a variable bit rate source and whether the service is connection or connectionless.

In order to support these different types of traffic, a different ATM Adaptation Layer (AAL) is defined to provide a service to each of these classes (AAL1–5, with AAL3 and 4 combined during a revision of the standards). AAL1 can be used to deliver 64 kbps voice through a constant bit rate service; AAL2 provides a variable bit rate service for example for compressed voice with silence detection (e.g. G.729a). AAL3 and 4 for connectionless data services and AAL5 is a refinement of the AAL3 and 4 specifications.

Why a 53-byte cell size and a 48-byte payload? The reasoning on size of cells was one of compromise between the data people who wanted a large cell size to carry more information in each transmission unit (let's face it the line rate is going to be many megabits per second, so it wouldn't take long to transmit a large cell and if you lost one, not a problem just retransmit it), and the voice people who wanted a small cell size. The voice people were looking to small cells to minimise latency in buffers on ports on the ATM switches and looking at reducing any problems associated with packet loss (the smaller the frame, the fewer voice samples per frame, the fewer samples lost if a frame gets dropped).

All of these cells then need to be multiplexed on to the appropriate physical layer (Synchronous Digital Hierarchy (SDH), Synchronous Optical NETwork (SONET), DWDM or even digital subscriber lines for example). This is the role of a transmission convergence (TC) sublayer.

We stated earlier that ATM is a connection-oriented service, so how are connections managed and for that matter created? The 5-byte header field on each ATM cell contains identifiers: Virtual Path Identifier (VPI) and Virtual Channel Identifier (VCI), the two combined representing a virtual circuit identifier and a virtual path is a collection of virtual channels going between two points. These fields represent the logical connections that the data in the payload (48 bytes) are being transported on. The connections can be pre-provisioned connections (Permanent Virtual Circuits (PVCs)) or on-demand connections (Switched Virtual Circuits (SVCs), although this term is out of favour). The important part of this nomenclature is the word virtual. Whilst a connection context exists using a particular combination of VPI and VCI fields, if the endpoint of the connection is not generation data, no cells are transmitted for this endpoint, unlike Time Division Multiplex (TDM) for example where, even if there is no speech timeslots are transmitted across the network.

The important point about the use of the VPI and VCI fields is that their significance is local to the switch node they are passing through, they do not have global significance unlike for example an IP address. Clearly globally significant VPI and VCI addresses wouldn't be possible; there just aren't enough bits in the ATM cell header. So this begs the question how is a connection across multiple switch nodes in an ATM system created? Figure 7.2 shows an example of the use of VPI and VCI fields. In fact the usage of VPI and VCI is not defined by the ATM standards and is left to the implementation.

ATM's Quality of Service (QoS) aspects include monitoring and control of such aspects as cell loss ratio, cell delay, and delay variance. QoS in ATM is defined as an end-to-end characteristic, the properties such as cell loss ratio are a measure based on this fact. The ATM forum, in a measure to simplify QoS for users, has defined five classes to express the service characteristics at the UNI. These are:

1 Class 0: No QoS guarantees, best effort service
2 Class 1: Constant Bit Rate applications (CBR) for example circuit emulation
3 Class 2: Variable Bit Rate (VBR) real-time traffic for example compressed packetised speech
4 Class 3: Connection oriented services
5 Class 4: Connectionless protocols

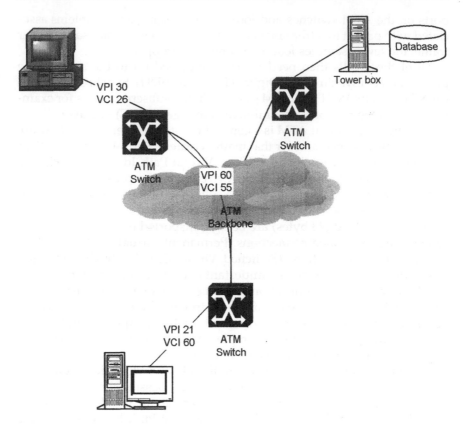

Figure 7.2   ATM VPI and VCI usage

It is left to the network operator to assign the values of cell loss ratio, cell delay, etc. to each of these classes and how best to manage these factors in their network. QoS is a complex topic and requires more consideration than is available in this text.

Clearly there is a lot more to both broadband ISDN and ATM than this short chapter covers, it is hoped this chapter gives a flavour of the power of ATM and its use for a multimedia carrier. Significantly more detail is covered in [BLACK2, DYSA] and are a recommended read if you want to (or need to) know more.

8

Representing Information

8.1 INTRODUCTION

Throughout computing history information has been represented in various forms from basic text through rich-text formats, postscript to binary encoding. More recently information and the way it should be presented with its roots in publishing has found its way to the top in the form of markup languages. The markup language most people will be familiar with (perhaps unknowingly) is Hypertext Markup Language (HTML). If you use a web browser, you're using HTML and in future its successor Extensible Hypertext Markup Language (XHTML).

The new kid on the block is Extensible Markup Language (XML) and it has found its way into almost every facet of telecommunications from provisioning services through to billing records and network management systems and even scripting languages for automated voice services.

The long history of telecommunications as a real-time systems design problem has in the past necessitated optimisations in the use and encoding of information. As hardware has got faster and more sophisticated, the need to encode information in protocols and databases in binary form is becoming less of an issue. The use of text encoding, borrowed from the Internet school of design because of its simplicity and ease of understanding, has become commonplace both in representing information in databases and in encoding protocol messages and remote procedure calls. It is this increase in capability, that combined with the view that content means revenue, is giving rise to the success of markup languages like those described in this chapter.

In this chapter we explore (X)HTML, XML and XML's children that have invaded the telecommunications network: Voice Extensible Markup Language (VoiceXML), Simple Object Access Protocol (SOAP), Universal

Description Discovery Integration (UDDI), Web Service Description Language (WSDL), Internet Protocol Detail Record (IDPR) and Call Processing Language (CPL).

Other notable content markup languages are the Wireless Markup Language (WML) made famous in Europe as part of the Wireless Application Protocol (WAP) standards, and the use of a cut down version of HTML in Japan for the i-mode data services. Neither WAP nor i-mode is covered here. i-Mode is essentially based on HTML which is covered. WAP is currently undergoing revision from its 1.1 specification to version 2.0. Version 2.0 of the specifications marks a dramatic change for the WML part of the specifications, which are now based on XHTML. The latest specifications for WAP can be found at http://www.wapforum.org and I refer the reader to [MANN, WAPF] for books on the topic.

In Part 2 we will explore the use of these technologies in services.

8.2 (X)HTML

HTML has done the World Wide Web (WWW) proud, so why change things and move up to XHTML. The simple answer is extensibility. HTML has proved difficult to extend (i.e. the addition of new markup components). HTML's history has allowed lax coding to take place and some tags don't need to be closed off in order for a web browser to correctly display the information. XHTML is much stricter over the coding rules. This is because of their family tree. HTML is defined as a Standard Generalised Markup Language Document Type Declaration (SGML DTD), whilst XHTML is defined using an XML DTD. XML has as its parent SGML. It is XML that imposes the stricter tag rules rather than SGML.

HTML is not a pure markup language and several developments have tried to address this, such as cascading style sheets. Without wanting to dive into XML, the difference between HTML and XML is that XML has no presentation information in the tags, it is purely a semantic definition language. HTML is a bit of a hybrid, with tags that describe how an element should be displayed (bold, italic and text colour for example). XHTML doesn't really fix this issue as it was designed with a degree of backward compatibility with HTML in mind.

What is the underlying reason for this change, XML parsers! More and more content is being parsed by programs other than browsers (for example in Business to Business (B2B) information exchanges). XML parsers find it difficult to parse HTML documents with missing tags (none well formed documents).

8.3 XML

Just about any book you pick up on XML will give you a brief history on XML's roots in publishing, so I won't bore you with that here; what I will say though is that XML is the natural extension to where the web was/is heading. How can I say that? In the previous section I noted the development of Cascading Style Sheets (CSSs) as being a factor in the development of XHTML. CSSs are a move to abstract the content of an HTML document from the formatting information; it is this approach that is at the heart of XML.

XML presents users of it with a powerful approach to representing the information they wish to communicate. It opens up so many possibilities from protocol specification to standard, human and machine-readable record formats. It is a true enabling technology.

Why is XML so powerful? Its specification creates a clear separation between semantic definition of the information and how it should be presented. This is extremely important and is arguably why XML has been chosen as the lingua franca of e-commerce and its choice for next-generation charging records (see IPDR later). The separation of content or meaning from how it should be displayed, means information need only be created once, it is then a matter of transforming (using XML style language transformations) the information into whatever is necessary to display it ((X)HTML, postscript, WML, etc.).

XML in its own right is not the important point! It is what you can do with XML that *is* important. In the following sections you will discover some of the uses XML has been put to, there are many more (in fact new ones are being created every day), but these are some of the important ones.

So what exactly is XML, as its name suggests it is a markup language, i.e. a way of using specific elements and attributes in a document so that it can be organised and stored in a constructive way. More specifically XML is even more powerful than this, because it can be used to specify the element in the first instant. For example consider a book (similar to this one), it consists of:

Front matter, the introductory section of the book consisting of
A half title page
Title
Title page
Title
Subtitle (optional)
Edition (optional)
Author
Publisher's imprint
Title verso
Copyright information

        Publishing history
      Contents
      Dedications
      Foreword
      Preface
      Acknowledgements
  A body
    Sections
      Part title
      Chapters
        Paragraphs
        Diagrams
  Back matter
    Bibliography
    References
    Index

Clearly from the way I have represented the book, it implies a structure, some parts contain other parts and the whole container is the book. So that everyone understands that a book looks like this structure and contains the elements above, a formal definition is needed. In XML terms this is called a Document Type Definition (DTD). In XML the elements can be thought of as representing a tree-like structure with the root at the start of the tree and the farthermost elements as leaves. The DTD for the book is shown below, combined with the xml for a document based on the DTD. The important point to take in is that the so-called root element is represented by the keyword 'DOCTYPE', so in this instance the root element is BOOK. The outmost elements (leaves) are paragraphs and diagrams, etc., so the sequence of element definition represents the tree structure. This example has now created a template that anyone can use to represent a book. An XML document using this template might look something like:

```
<?xml version="1.0" standalone="yes" ?>
<!DOCTYPE BOOK [
  <!ELEMENT BOOK ANY>
  <!ATTLIST BOOK isbn CDATA "">
  <!ATTLIST BOOK price CDATA "">
  <!ELEMENT HALFTITLE (HALFTITLE_TITLE)>
  <!ELEMENT HALFTITLE_TITLE (#PCDATA)>
  <!ELEMENT TITLEPAGE (TITLEPAGE_TITLE, SUBTITLE*,
            EDITION*, AUTHOR, PUBIMPRINT)>
  <!ELEMENT TITLEPAGE_TITLE (#PCDATA)>
  <!ELEMENT SUBTITLE (#PCDATA)>
  <!ELEMENT EDITION (#PCDATA)>
  <!ELEMENT AUTHOR (#PCDATA)>
  <!ELEMENT PUBIMPRINT (#PCDATA)>
  <!ELEMENT TITLEVERSO (COPYRIGHT, HISTORY)>
```

```
<!ELEMENT COPYRIGHT (#PCDATA)>
<!ELEMENT HISTORY (#PCDATA)>
<!ELEMENT CONTENTS (#PCDATA)>
<!ELEMENT DEDICATIONS (#PCDATA)>
<!ELEMENT FOREWORD (#PCDATA)>
<!ELEMENT PREFACE (#PCDATA)>
<!ELEMENT ACKNOW (#PCDATA)>
<!ELEMENT BODY (SECTION + )>
<!ELEMENT SECTION (PARTTITLE, CHAPTER + )>
<!ELEMENT PARTTITLE (#PCDATA)>
<!ELEMENT CHAPTER (PARAGRAPH + , DIAGRAM*)>
<!ELEMENT PARAGRAPH (#PCDATA)>
<!ELEMENT DIAGRAM (#PCDATA)>
<!ELEMENT BACKMATTER (BIBLIO, REFERENCES, INDEX)>
<!ELEMENT BIBLIO (#PCDATA)>
<!ELEMENT REFERENCES (#PCDATA)>
<!ELEMENT INDEX (#PCDATA)>
]>
<!-- My Book -->
<BOOK isbn="1-11235-661-1" price="£40">
  <HALFTITLE>
    <HALFTITLE_TITLE>This is the half title page title of my
        book</HALFTITLE_TITLE>
  </HALFTITLE>
  <TITLEPAGE>
    <TITLEPAGE_TITLE>This is the title page title of my book
    </TITLEPAGE_TITLE>
    <EDITION>This is the second edition of the book</EDITION>
    <AUTHOR>Neill Wilkinson</AUTHOR>
    <PUBIMPRINT>This is the text for the publisher's imprint
        </PUBIMPRINT>
  </TITLEPAGE>
  <TITLEVERSO>
    <COPYRIGHT>Copyright text</COPYRIGHT>
    <HISTORY>This is the publishing history of the book
        </HISTORY>
  </TITLEVERSO>
  <CONTENTS>The contents pages of the book</CONTENTS>
  <DEDICATIONS>Dedications</DEDICATIONS>
  <FOREWORD>Someone has written something really inspiring about
        the book. </FOREWORD>
  <PREFACE>Why did I write this book?</PREFACE>
  <ACKNOW>I'd like to thank everyone!</ACKNOW>
```

```
<BODY>
  <SECTION>
     <PARTTITLE>Title of the section</PARTTITLE>
     <CHAPTER>
       <PARAGRAPH>First Chapter paragraph</PARAGRAPH>
     </CHAPTER>
     <CHAPTER>
       <PARAGRAPH>Second Chapter paragraph</PARAGRAPH>
       <PARAGRAPH>Blah Blah</PARAGRAPH>
     </CHAPTER>
     <CHAPTER>
       <PARAGRAPH>and on Chapter</PARAGRAPH>
       <DIAGRAM>This is where a diagram goes.</DIAGRAM>
     </CHAPTER>
  </SECTION>
  <SECTION>
     <PARTTITLE>Title of the second section</PARTTITLE>
     <CHAPTER>
       <PARAGRAPH>First Chapter paragraph Section 2
         </PARAGRAPH>
     </CHAPTER>
     <CHAPTER>
       <PARAGRAPH>Second Chapter paragraph Section 2
         </PARAGRAPH>
       <PARAGRAPH>Blah Blah</PARAGRAPH>
     </CHAPTER>
     <CHAPTER>
       <PARAGRAPH>and on Chapter 2 Section 2</PARAGRAPH>
       <DIAGRAM>This is where a diagram goes.</DIAGRAM>
     </CHAPTER>
  </SECTION>
</BODY>
<BACKMATTER>
  <BIBLIO>Some kind of bibliography</BIBLIO>
  <REFERENCES>Lots of references to see how much research was
       done</REFERENCES>
  <INDEX>A list of all the keywords plus page numbers
         </INDEX>
</BACKMATTER>
</BOOK>
```

The `standalone="yes"` says that the DTD and document are together in the same file. Elements (tags) can also have attributes. This is defined in the `ATTLIST` line in the DTD. The distinction of when to use attributes and when to use sub-element is far from clear and only guidelines are given in the standards, but essentially it is left to the designer of the XML document. One final note I have used capitalised elements and lower case attributes, this is purely a design decision I made when constructing the example and mixed case is valid for both elements (tags) and attributes. The use of DTDs will no doubt be replaced by 'schemas', a recently standardised XML document with more power than the DTD and a common XML structure than the DTD structure.

Hopefully this simple example should explain how XML is used and if you look up some of the specifications for VoiceXML, WML, etc. then they should make a little more sense. I recommend you look at any of the books available on XML, it is a well-written about topic and there are plenty of references.

8.4 VOICEXML

VoiceXML is an incentive to standardise the way applications for media servers (Interactive Voice Response servers (IVRs)) are written, the current way applications for voice platforms are scripted is in a vendor proprietary manner. This means applications for one platform can't readily be transported to another. The VoiceXML forum are keen to promote the use of VoiceXML, the main reason being it's easier to develop VoiceXML applications than proprietary applications and there will be more web savvy developers than specific developers trained in a proprietary IVR scripting language, thus bringing down cost.

The other thing that is promoting the use of VoiceXML is the increase in processing power and the improvements in Digital Signal Processing (DSP) that are finally making voice recognition a reality and text-to-speech more natural. VoiceXML is not completely reliant on these technologies, as we will see later, dialogues can be constructed from pre-recorded prompts and use DTMF selections. VoiceXML also builds on the web model for delivery of content and is structured around Hypertext Transfer Protocol (HTTP) and web servers for content delivery.

VoiceXML is an XML document that allows the structuring of complete applications that allow integration with back office web services for the retrieval of content and the posting of information. The vxml document contains tags for the construction of: 'dialogues', 'forms' and 'menus'. Forms and menus are the two types of dialogue, with a menu being used to construct a flow of control based on a choice. Dialogues are broken up into either field items ('field', 'record', 'transfer', 'object', 'subdialog' tags) or control items ('block' and 'initial' tags).

Within a dialogue, tags for prompting the caller with voice can use one of: text-to-speech ('prompt' tag), pre-recorded audio files ('audio' tag) or streaming audio files ('audio' tag plus caching and fetchhint attributes set to not cache the file and to start playing it before it has been completely retrieved); tags for getting caller input are also specified ('field' tag).

The basic structure of a VoiceXML document is as follows:

```
<?xml version="1.0"?>
<vxml application="my_first_vxml_app" version=
        "1.0">
<form id="first_form">
  <field name="field1">
    <block>
    <prompt>
      Hello World
      <audio src="www.telecomsoapbox.org.uk/
          hello.wav">
    </prompt>
    </block>
  </field>
</form>
<form id="GetInput">
<!-- This field will collect up to 15 DTMF digits -->
  <field name="Input"
          type="digits?minlength=1;maxlength=15">
    <prompt> Give me DTMF input </prompt>
  </field>
</form>
</vxml>
```

Control of flow between forms is also possible, clearly necessary if you are going to give callers choices in the form of menus, via the 'goto' tag and 'if', 'else' and 'elseif' tags.

Clearly as a complex application with multiple paths starts to be developed it would make sense to break the documents up into smaller more manageable chunks. These could then be given to different developers to code up. This facility is provided by specifying the document uri in the goto tag `<goto next="next_document.wxml">`.

OK I think that's quite enough VoiceXML to keep anyone happy. Hopefully this brief coverage has given you enough to get you into the feel of what VoiceXML is capable of. It can do pretty much the same as HTML in linking documents together across the web. One note of caution about linking documents and retrieving large voice files, make sure you've got the bandwidth unlike a web browser interface, people will not hang around in a voice enabled system waiting for the next prompt to load! If you want to know more about VoiceXML, then I suggest a visit to the VoiceXML forum's website (http://www.voicexml.org).

8.5 SOAP, UDDI AND WSDL

Distributed computing has evolved in the last decade around technologies such as the Common Object Request Broker Architecture (CORBA) and other inter-process communications mechanisms, such as Microsoft's Distributed Common Object Model (DCOM) (see [ORFA] for a good coverage of these topics). Web content distribution has evolved around the WWW and technologies such as HTTP and latterly XML. The distributed computing camp brought to the table mechanisms for objects and applications to interact and discover the methods that other objects use to perform their tasks. The web brought content representation and open communications to the table. The result of this marriage is Simple Object Access Protocol (SOAP), Universal Description, Discovery and Integration (UDDI) and Web Service Description Language (WSDL). These techniques are currently at the forefront of the move to create distributed web services.

SOAP and WSDL have grown out of the work on Microsoft's.NET distributed application framework and other work in Compaq, IBM, HP et al. and have been submitted to the World Wide Web consortium (W3C) for standardisation.

An independent consortium runs the UDDI work (www.uddi.org) that aims to enable businesses and services to discover each other and define how they can interact in an open, platform-independent way through the use of a global registry. This work, the UDDI group claim, will be handed over to a standards body.

These techniques combined will enable dynamic B2B transaction to take place and much more, the opportunity for the creation of services from either off-the-shelf components, pre-build by third parties or actually in existence on the Internet could become a reality that the object-oriented community has been striving for.

So let's take each in turn and have a brief look at what they do. SOAP is a mechanism for encoding information for exchange between two applications and for the encoding of procedure calls or object methods in an XML document and exchanging them over a transport protocol. The transport protocol originally specified for SOAP is HTTP, however, work has been done (in the form of an Internet draft) to transport SOAP over Session Initiation Protocol (SIP). The standardisation effort is taking place within the XML protocol area in the W3C.

SOAP has three parts: an envelope that describes the message and what is needed to process it, a collection of encoding rules that describe how data types are defined (for example a C program has char, int, short, etc. data types, these need to be represented in any protocol exchange) and finally a convention that states how remote procedure calls and responses should be formatted (Figure 8.1).

So a SOAP message is an XML document that consists of an envelope (mandatory), an optional SOAP header and mandatory SOAP body. The body is in effect the contents of the envelope and contains the information intended for the recipient of the message.

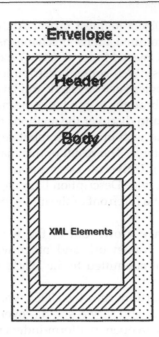

Figure 8.1   SOAP structure

The body contains an XML element that can represent function calls and data items.

WSDL defines an XML grammar (DTD) for the definition of communications services. The services are described as a collection of endpoints or ports that are capable of exchanging messages. The whole package is bundled as a set of definitions. Each definition is a collection of:

- type – data types that are used by the service
- message – this is a list of the data that are being transferred, typed by the data types previously defined
- port type – essentially the definition of one or more functions or object methods including the function call parameters, the parameters being defined in the message portion
- a port – this defines the address at which the function calls can be made
- service – a collection of ports
- documentation – a free format text area for human readable documentation

OK so that's all pretty abstract – what does that actually mean. Imagine a function call in a procedural language like C, the definition might be:

```
int getTime (int offsetfromGMT);
```

A call to this function might look like:

```
int Australian_time, offset;
offset=11;
```

```
Australian_time=getTime (offset);
```
The hope is that the function call returns a value into the variable `Australian_time` that is Greenwich Mean Time plus 11 h. Now that's fine if the function is part of the same program, but what if the function was part of a service that was out on the Internet somewhere. Firstly how would you know what the function definition was if you wanted to call the function, let alone where the service is that implements time offsets. How would you encode the function call on the 'wire'? Let's have a look at an example WSDL definition for this service (http://telecomsoapbox.-org.uk/time/timefunction.wsdl):

```
<?xml version "1.0" ?>
<definitions name= "GMTTimeOffSet"
targetNamespace= "http://telecomsoapbox.org.uk/time/
        definitions"
xmlns:tns= "http://telecomsoapbox.org.uk/time/
        definitions"
xmlns:xsd1= "http://telecomsoapbox.org.uk/time/
        definitions"
  :
  : other namespace definitions>
<import namespace= "http://telecomsoapbox.org.uk/
        time/schemas"
location= "http://telecomsoapbox.org.uk/time/
        schemas/time.xsd" />
<documentation>
This is where the function parameters are defined
</documentation>
<message name= "Australian_time" >
  <part name= "body" element= "xsd1:
        getTimeReturnType" />
</message>
<message name= "offsetfromGMT" >
  <part name= "body" element= "xsd1:
        getTimeoffsetType" />
</message>
<documentation> Now the function definition
        </documentation>
<portType name= "getTimePortType" >
  <operation name= "getTime" >
    <input message= "tns:offsetfromGMT" />
    <output message= "tns:Australian_time" />
  </operation>
 </portType>
</definitions>
```
The stuff at the top of this XML document is standard XML information

about the document type, i.e. XML 1.0, and then we have the tags for the WSDL document. The first elements define the namespace of the document; you'll also notice that the document imports a schema definition from time.xsd. This defines the types of the message elements, and is used to illustrate the fact that you can separate out parts of a WSDL document to aid clarity.

Then follows the definition of the function call parameters, followed by the definition of the function call itself. So that completes the definition of the function call, but not the service or where it resides. In order to make the service concrete and allow the function call to be mapped to some form of protocol a process called 'binding' is used. This process maps a protocol to the abstract definition of the portType by way of a port and to a service by way of the 'service' tag. The bindings described for WSDL include SOAP, so SOAP can be used as the protocol for exchanging messages.

For example, the time conversion function above the SOAP bound service might look like (partial document):

```
<definitions>
  <import namespace="http://telecomsoapbox.org.uk/
       time/timefunction.wsdl"/>
  :
  :
  <service name="GMTtimeconversion">
  <port name="timeconversion" binding="tns:gettime
       binding">
     <soap: address location="http://telecomsoapbox.
       org.uk/time"/>
  </port>
  </service>
</definitions>
```

If you want to study SOAP or WSDL further then I recommend you pay a visit to the W3C's website at http://www.w3c.org, where you will find more examples and the specification documents for SOAP and WSDL, or look at [SCRIB] for more on writing applications with SOAP and look at http://www.uddi.org for the work on UDDI.

8.6 IPDR

Current call records, media, rating and billing systems are a complex collection of integrated IT systems. Some of the progress of billing and arguably the cost burden is derived from the lack of a common format and means of delivery switch records that record call usage (just about every switch manufacturer has different billing record format!). These records have to be collected, sequenced and transformed through mediation systems before rating engines can process them and finally delivered to the systems that produce customer bills. This problem can only get worse (and is getting worse) as the telecoms industry is moving towards a more content driven future, where the value in the network is actually the content, not the resources it requires to deliver that content, just as Plain Old Telephony Service (POTS) based telephony services have become commodity services and suffered price erosion in the face of competition, so will the eventual delivery of the broadband network. We are already seeing residential broadband services being offered at relatively low cost. How you represent the transactions for these future services is far from trivial, the IPDR.org site gives the example of a simple email transaction containing components such as size, time of day, delivery options, etc.

The IPDR.org group is looking to provide a standard-based approach to the exchange of information between network elements, Operational Support Systems (OSS) and Business Support Systems (BSS). The main document release of IPDR.org is the 'Network Data Management – Usage (NDM-U) for IP-based Services', that describes a framework that describes the relationship between Internet Protocol (IP) network and service elements and support systems and the information flows between them. It also defines an XML schema that specifies a protocol for exchange of information (based on SOAP) between elements and an XML schema of the usage attributes defined by the IPDR.org group for these potential services.

I am not specifically endorsing the work of this group here (plenty of others are doing that already – the Global Billing Association, the International Engineering Consortium and the TeleManagement Forum), more highlighting that one this work is taking place and that two this work is necessary if the industry is to move forward with a means of reducing cost of service provision and billing an increasingly competitive and price conscious market. If you would like to know more on the work of the IPDR.org then I suggest you check out http://www.ipdr.org.

The important point about the use of XML in the definition of session detail records is XML's power to represent data, without presentation information, combined with XSLT's ability to transform one XML document into another, means mediation devices will be able to be constructed much more easily (potentially reducing both the time and cost associated with the introduction of new devices into a telecoms network).

8.7   CALL PROCESSING LANGUAGE

Like VoiceXML, Call Processing Language (CPL) is an XML defined scripting language. It is at the time of writing still a draft standard in the IP telephony working group of the Internet Engineering Task Force (IETF), but will no doubt be ratified by the time this is read.

The idea behind CPL is to enable the easy implementation of user configured services that can run on call servers (see Section 5.6 for a description of call servers). Whilst CPL has been used with SIP so far,[1] it is a protocol independent language. CPL can also support client application scripting to support the writing of personal script to control for example SIP clients.

Like all XML documents it has a document type definition. In CPL's case this is defined in draft-ietf-iptel-cpl-04.txt (the latest version available at the time of writing).

The specification of CPL defines a set of primitives that form a simple one-way tree structure. This tree structure represents the decisions that make up the service. The language primitives are:

- Switch nodes: these are the decision points in the script that allow different actions to be taken. Decisions can be made based on elements of the message that triggered the script to run (such as sender field, recipient field, etc.). Other decisions such as time and date are also allowed. This can be used to create time of day sensitive routing scripts.
- Location nodes: these specify the location signalling actions should be directed to.
- Signalling actions: signalling actions are what control the behaviour of the script and in essence allow the script to do something useful. There are three actions: proxy, redirect and response (you can see the SIP heritage here). The proxy action causes the server running the script to send the received message that triggered the script to the currently specified location held in the location node. The script then waits for a response. Redirect causes the server running the script to send the message on to the location set in the location node and terminate the script. Finally respond creates a response (failure or rejection of the call) and the script exits.
- Non-signalling actions: these are mechanisms for the script to record events say in a system's log, or send email or instant messages to a user.

Each of the nodes described above is represented in CPL as a pair of tags. The example below shows a CPL script that causes a call to be redirected to two different places (office or home) based on a decision on the time of day. For unsociable hours route all calls to voicemail.

```
<?xml version="1.0" ?>
<!DOCTYPE cpl PUBLIC "-//IETF//DTD RFCxxx CPL 1.0//EN"
        "cpl.dtd">
<cpl>
<incoming>
<timeswitch>
```

---

[1] Dynamicsoft support CPL scripting on their SIP application server.

```
<!--Redirect calls starting 1 Jan 2001, 08:30 for 9 hour
        Mon - Fri-->
<!--to office -->
  <time dtstart "20010101T083000" duration "PT9H"
        byday "MO,TU,WE,TH,FR">
    <location url="sip:office@company.com">
        <redirect />
    </location>
  </time>
<!--Redirect calls starting 1 Jan 2001, 17:30 from
        5 Hours -->
<!--to home -->
  <time dtstart "20010101T173000" duration "PT5H"
        byday "MO,TU,WE,TH,FR" >
    <location url="sip:home@myisp.com">
        <redirect />
    </location>
  </time>
<!--Redirect calls to voice mail-->
  <otherwise>
    <location url="sip:myvoicemail@telco.com">
        <redirect />
    </location>
</timeswitch>
</incoming>
</cpl>
```

9

Directories – More Than Just Information Storage

9.1 Introduction

Directories are rapidly becoming the cornerstone to delivering flexibility for advanced services, both as means of storing location information, user configuration and resource privileges and as a means of storing service-specific information allowing for rapid provisioning of customers and rapid modification of services. One of the key issues that directories will resolve for the next-generation networks is the dynamic nature of the future services and the desire to reach services based on useful text-based naming.

In the previous chapter on representing information we explored a number of exciting uses of Extensible Markup Language (XML) to allow applications to discover and use services, none of this discovery process would work without directories.

I start this chapter with a brief look at the domain name system (DNS), possibly the best-known directory system, since most of us use it every day. I continue the chapter by looking at X.500 and Lightweight Directory Access Protocol (LDAP). X.500 is the International Telecommunications Union (ITU) standard for directory technologies and whilst complex is arguably the most influential of all the directory technologies. We then take a look at the concept of meta-directories, directory systems that link other directories together.

9.2 Domain Name System (DNS)

The DNS was created to overcome the scaling problems associated with the distribution of a text file (hosts.txt) based database that listed all the hostnames and their Internet protocol (IP) addresses.

The main part of the DNS is the definition of a hierarchical naming scheme based on a domain model and a distributed database (directory). Two application components are used to interrogate the database, a 'resolver' for the client application to use and a server application called the 'domain name server' that accesses the directory and responds to the query.

The hierarchical naming scheme in the DNS breaks the address space into an upside-down tree structure, with several hundred Top-Level Domains (TLDs) sitting at the top of the tree, with one single un-named root. The structure of the tree is reflected in the namespace by the use of periods (.) that separate branches and leaf nodes (branches and leaves being known as labels). For example www.telecomsoapbox.org.uk, www is the leaf, telecomsoapbox a branch of the org branch, which itself is a branch of the TLD uk. Figure 9.1 shows this represented as an upside-down tree.

We mentioned above that the database is distributed; this is achieved by virtue of the hierarchical address space. For example I own the domain telecomsoapbox.org.uk. If I chose to run my own domain name server for this domain, I could choose any name I liked within this namespace, for example my PC on my desk has the name neillspc (original I know!) and in the telecomsoapbox.org.uk domain it is neillspc.telecomsoapbox.org.uk.

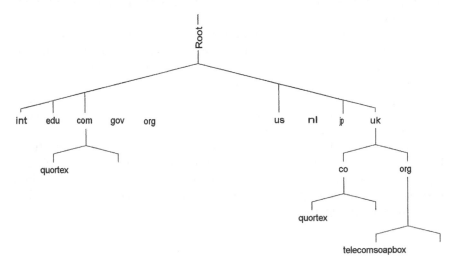

Figure 9.1 DNS name space tree

OK so that's fine and dandy, I can create my own names for my own addresses, but how does anyone find these out or for that matter actually contact the host which, whilst the name is useful to humans, computers like to refer to computers on the Internet by their IP address. That's where the resolver and the domain name server come in. When an application (say a web browser) gives the resolver a name, www.telecomsoapbox.org.uk for instance, what the resolver does is issue a dns request to the domain name server giving it the name, the server responds with the 'resource records' for that name. Now in most cases that is just an IP address, but other resource records exist (start of authority – SOA, mail exchange – MX, name server – NS, etc.).

The last paragraph explained the records and that a name server is responsible for delivering the results of the query, but clearly you need to be able to connect to a name server in the first place and since we've already said that the DNS is distributed and each domain can run their own server, then there's clearly more than one. These multiple servers are grouped into 'zones', with each zone having an authoritative name server (essentially the root of that zone). When a resolver needs to query a name server, it queries its local one, which is usually configured in the host either manually or via dynamic allocation using for example Dynamic Host Configuration Protocol (DHCP). The local domain server may not know anything about the name it has been given, so it sends the query on to the TLD server for the domain requested. If the top-level server does not know the address, it must refer it to one of its children and so on until the server that is the authority for that domain is found. The response finds its way back down the chain.

The alternative to this chaining mechanism is for the name server to respond with the address of another name server that may know the answer (referral). In this case the resolver must try this address instead and so on. Also clearly it is inefficient to keep repeating a long path to an authoritative source, so most domain name servers keep a cache of addresses with time to live fields so that they know when to flush the cache.

That's a very brief encounter with the DNS, the salient point is that the DNS is what runs the name resolution mechanism of the Internet. The records held by the domain name servers are added to over time to be able to resolve different entities for example the use of a SRV record (RFC 2782) has been proposed to include the identity of Session Initiation Protocol (SIP) servers for a particular domain. The most recent development around the DNS is under the ENUM working group in the Internet Engineering Task Force (IETF). The working group is looking to add telephone number records to the DNS database that will allow translation to for example Uniform Resource Locators (URLs). The idea is that a telephone number might be used to reference a number of services.

9.3   X.500 and LDAP

X.500 is a collection of recommendations for the organisation, storage and publication of directory information. The main downside, some might say, about X.500 is its size, complexity and reliance on the OSI protocol stack. To a degree all this is true, but with the use of the recommendations in RFC 1006 (OSI applications over IP) and a cut down 'light' version of the directory access protocol (LDAP), X.500 directories represent powerful, accessible repositories of information.

OK the basics; like the DNS, X.500 specifies a structure to the information in the directory in a hierarchical way, this is called the Directory Information Tree (DIT). The top of the tree is referred to as the 'root', branches and leaves are referred to as 'containers' and there is also a pointer or 'alias' element, which allows referrals to other leaves of the DIT. A DIT that has been populated with information is known as a Directory Information Base (DIB).

The directory server(s) is organised into authorities or organisations, known as the Directory Management Domain (DMD). Each directory server is known as a Directory Systems Agent (DSA) and the equivalent client application to the DNS resolver is called a Directory User Agent (DUA).

DUAs interrogate directories through DSAs and the three mechanisms for finding information are chaining, referral and multicast. The first two are the same as the operations described for the DNS, multicast is where the DSA sends a request to multiple other DSAs either in parallel or sequentially.

The way a DUA communicates to a DSA is through the Directory Access protocol (DAP), the original DAP communicated through the OSI stack, which made for a complex DUA that was difficult to implement on early PCs. The IETF RFC 1006 combined with a light version of the DAP – LDAP. LDAP version 3 is specified in RFC 2251 and is supported by RFC 2252 which describes the syntax of the attributes of the directory to help define the directory schema for use with LDAP and RFC 2849, which describes the LDAP Data Interchange Format (LDIF) used to import and export information from an LDAP directory.

Unlike the DNS which is organised around the resolution of servers and server addresses, X.500 is organised around geographic, organisational and people elements. This has been found to be restrictive and alternative schemas have been proposed including incorporating the DNS scheme into an X.500 directory.

That was also a whistle-stop tour of X.500 and doesn't cover all the aspects of X.500, but hopefully gives a flavour of what it is about. XML has also got a place in defining a schema for an X.500 directory server, DirXML has been proposed and also Directory Services Markup Language (DSML – see www.dsml.org) as a means of mapping directory information obtained from LDAP queries into an XML document. The idea behind this is to allow easy access to multiple directory sources through a standard specification. The ideas of DSML bring us nicely to meta-directories....

9.4   The Meta-Directory

A meta-directory simply put is a consolidation point for information from multiple directory sources and is proposed to link organisational (internal) directories to external directories. The idea is to provide a single point solution for application to access that can transparently synchronise the update of information from multiple data sources. A meta-directory is as much a data management process as it is a directory in its own right, and relies heavily on technologies such as LDAP and DSML to provide the glue to consolidate the multiple data sources.

Meta-directories will become increasingly important in the next-generation network. As applications get more complex and more intelligent, the need to synchronise information across multi data sources such as: provisioning databases, authentication, profiles, enterprise databases, service capability database (such as Home Location Registers (HLRs) and Service Data Points (SDPs)) and operational support and business support systems, will increase. Not to mention finding services that are out there. In Chapter 8, a discussion of Universal Description Discovery and Integration (UDDI) was put forward under the topic of representing information. UDDI is more than just this and represents a cornerstone in the web services infrastructure as a meta-directory of services, and how to find and communicate with them.

9.5   Other directory technologies and ideas

There is currently a proliferation of directory technologies that this short chapter cannot hope to cover, listed below are just some these:

- Microsoft's Active Directory. This is used in Microsoft networks from Windows 2000 onwards as a distributed repository of information about the users, applications and devices present on the network.
- NetWare NDS. This provides a distributed system similar to that of Active Directory.
- The Common Information Model (CIM), network management and the Directory Enabled Network (DEN) proposals.
- Service data points and HLRs (see Chapter 4).
- Remote Address Dial-In User Service (RADIUS) server. This is used to control admission to dial-up network resources.

I hope this chapter has given a taste of directories and their importance in the next-generation networks. I suggest for a more complete explanation of directories and their technologies the reader looks up [KAMP].

Part II:
Services, Architectures and Applications

INTRODUCTION

This section is where we explore some of the services that will be present in the next-generation networks. This is achieved through the examination of how existing services may evolve to meet the needs of the next-generation networks, by putting all the technologies outlined in the first section to good use and exploring the architectures of these new networks.

Firstly we'll look at Intelligent Networks, the cornerstone to a number of advanced services in the current circuit switched PSTN and look at how existing services work and how they might change. We'll then take a look at Interactive Voice Response (IVR) and how the IVR has managed to re-invent itself in a next-generation network with the use of VoiceXML to open up its scripting engine and its place in Unified Communications servers.

Call Centres have long been the standard bearer of convergent applications, bridging the divide between enterprise database systems and intelligent call control. We'll examine the current architecture for call centres and postulate a future framework for call centre services using the Session Initiation Protocol (SIP).

The service provider market space is currently in a state of turmoil as incumbent telecoms operators, Internet Service Providers (ISPs) and new entrants all flock to provide customers with value added services hosted in their networks. Application Service Providers (ASPs), Application

Infrastructure Providers (AIPs) and other Service Providers will be explored and a view of what a next-generation hosted architecture might look like will be presented together with how these service providers could deliver their services to the customers.

We'll bring it all together into a total architecture encompassing fixed and mobile telephony with hosted services. We'll do this by way of application frameworks, an area of great interest for the software architecture of new service platforms.

10

Intelligent Network Services

10.1 INTRODUCTION

In Part 1 of this book, we examined the functional and physical characteristics of circuit switched based Intelligent Networks (INs). In this chapter,
we are going to explore what these elements do by way of offering
services to customers and giving carriers a flexible means of delivering
new services.

The IN service model was the first step to releasing service control from
the hands of switch manufacturers and as such presented telecoms operators with a new vehicle for realising services that enhanced the basic call
control capabilities of stored program control switches. We see that this is
not the end of the story for enhanced service platforms as the move to
remove the final block to enhanced services, the close coupling of the
stored program controller and the switch fabric, to release the potential
of software driven services on packet networks.

Just to refresh the reader on INs. They rely on the decoupling of a
number of telephone switch functions from the stored program controller,
renamed a Service Switch Point (SSP) in the IN architecture. The functions
left behind in the SSP are called the Basic Call State Model (BCSM). The
functions separated out are incorporated into a centralised service execution environment as a Service Independent Building Block (SIB) called the
Basic Call Processing (BCP) function. The services, constructed from SIBs
to form a Service Logic Program (SLP), run inside this environment called
the Service Logic Execution Environment (SLEE), inside a physical platform called the Service Control Point (SCP).

All of this distributed processing is facilitated by a protocol called

Signalling System Number 7 (SS#7). The remote method invocation protocol that runs on top of SS#7 is known as the Intelligent Network Application Part (INAP). Mobile telephone networks implement a protocol on top of SS#7, called the Mobile Application Part (MAP). This protocol is the glue that allows mobile networks to support roaming subscribers.

The combination of IN and mobile networks come together to deliver the Wireless Intelligent Network (WIN).

All the above is a brief recap of the information covered in Chapters 1–4.

10.2   EXAMPLE EXISTING SERVICES AND HOW THEY WORK

Arguably the most common service offered from an IN in the fixed network is basic number translation services. In fact it could be said that it was these services that created the desire for INs. The earliest launches of IN service in the US in the late 1980s, were by AT&T. AT&T used a centralised database connected to the telephone switches via an SS#7 network, which allowed the switches to request translations for 800 service numbers.

So how does such a service operate? Remember back to Chapter 1 on how the public switched telephone network is structured into a hierarchy of local, transit and international exchanges (switching layers). These layers are structured around the E.164 numbering plan with numbers having local, national and international significance. So when a new numbering range like the free phone services of 800 and local charge rate numbers such as 0345 (in the UK) are dialled, what is a local exchange supposed to do, they have no national or international significance with respect to the E.164 numbering plan.

The answer is that the routing tables in the circuit switches are configured to route the call to an additional layer in the network, the SSP layer. In some implementations this layer is combined with the trunk/transit layer and in others the individual local exchanges have been converted to SSP capable switches.

By way of an example, let us consider a layer of SSPs above the local exchanges combining the functionality of transit exchanges and SSPs. In this case the local exchange signals the called number up to the transit layer (via SS#7 Integrated Services User Part – ISUP), the combined transit/SSP has a trigger point set for interruption of the basic call routing software (BCSM). This causes the SSP to initiate a 'dialogue' with the SCP by sending an INAP initial-DP (Detection Point) message containing (amongst other items) the calling party's number, the called number (the 800 or 0345 number), a service key (to identify the service logic program to execute) and the event type in the BCSM that triggered the request. The call routing engine in the SSP then suspends the call processing, waiting for a response from the SCP.

The SCP, using the service key, executes the SLP associated with that key. The most common number translation SLP applications are:

- time of day routing, based on the time of day a call can be routed to a number of destinations, for example a call centre application might want all calls after 10 p.m. to terminate on a recorded announcement;
- geographic origin, the call can be connected to a number of different numbers based on the E.164 national prefix of the originating caller's number;
- proportional distribution of calls to different destinations, again for a customer with multiple call centres the calls can be evenly distributed amongst the different bureaux; multiple choice routing, for example for a find-me follow-me service the call may be directed to different numbers and based on the response (busy say) the call can be redirected until an answer is received or the caller can be diverted to voicemail;
- combinations of the above!

Once the service logic has determined the destination of the call, in the number translation scenario the SLP responds to the initial-DP message in one of two ways: Establish a Temporary Connection (ETC) or a simple CONnect (CON) procedure. Both ETC and CON contain the destination number to be connected to. The difference between the two methods of connection is that an ETC is still under the control of the SLP and is usually used to temporarily connect a caller to a recorded announcement or automated announcement service running on an Interactive Voice Response (IVR, called a Voice Response Unit (VRU) in the US) platform in the form of an intelligent peripheral or service node whilst the CONnect method completes the connection without further action.

Additional information can be requested from the SSP in the form of reports about for example charging. When the call completes, the SSP will send the SCP the report requested and the SLP invoked will return to an idle state.

What other services can make use of this kind of control, a whole host of them:

- Tele-voting, e.g. when a game show asks viewers to call in to register a vote for a contestant. The IN can automatically count calls to a specific number and the results can be obtained in real time.
- Virtual private networks. In a multinational company the connection of its private telephone network via the public telephone network can reduce the cost of routing calls by not having to lease expensive circuits between sites. IN can be used to add routing plans that adapt to time of day tariff changes for example.

- Wide area Centrex. Instead of a company connecting a private network together, the telephone company can create a network for them, including unique numbering plans. All of the intelligence and configuration is done on the telephone company's switches and IN systems.
- Calling card services. Most telecoms network operators around the world (and a number of partner companies) offer phone card services that allow people to charge calls to a separate account irrespective of where the call is made from. This normally involves calling a free number, which connects the caller first to an IVR. The customer then keys in account and personal identification codes, followed by the number they wish to reach. The IVR system then connects the call. All of this is generally performed using an IN system.
- Mobile networks have also made extensive use of IN for pre-paid mobile services. Pre-paid phones are validated by a service running on an SCP. If the caller has credit, then calls are allowed. Rules are set to automatically divert calls to an automated service when a particular threshold is reached on their credit.

In order for the SLP to perform its role in these services it utilises the services of a database in the form of the Service Data Point (SDP). The SDP contains the customer specific data that allows a specific SLP to be executed for many different customer instances. In real-world implementations the SDP is sometimes embedded in the SCP, rather than being a separate physical entity.

10.3 SOFTSWITCHES AND APPLICATION SERVERS

Just as the telecoms world was getting comfortable with the evolution from electromechanical exchanges, through common control and stored program control circuit switching and INs, in comes a fourth generation of servers – softswitches. Ovum is predicting the softswitch market to be very lucrative,[1] around US$1.7 billion in market worth by 2006 and application server market around US$1.1 billion. So it would appear that it is a market worth taking notice of.

The term softswitch is used to describe a server platform that is capable of controlling telephony and other services by the construction of programs in an Internet Protocol (IP) network. Ongoing work within the softswitch consortium is looking to clarify the architecture and interoperability of different vendors' softswitch products. The title softswitch can also be used to describe a Media Gateway Controller (MGC), a call agent, a SIP proxy server and a H.323 gatekeeper.

[1] Softswitches: the keys to the next-generation IP network opportunity

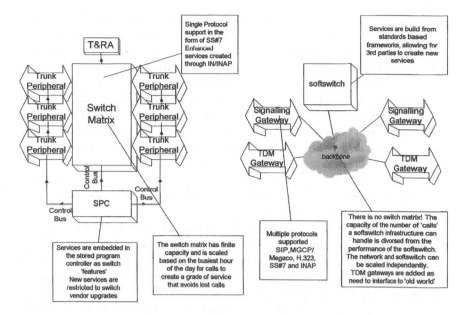

Figure 10.1 Stored program control and softswitch architecture

A softswitch goes one step further than each of these platforms to incorporate a service development environment built on top of a vendor neutral platform. Some vendors' products also come pre-configured to incorporate the common feature found in current stored program control based circuit switches.

The softswitch consortium's definition of a softswitch (aka call agent, call server or media gateway controller) is a device that provides, at a minimum:[2]

- intelligence that controls connection services for a media gateway, and/or native IP endpoints;
- the ability to select processes that can be applied to a call;
- routing for a call within the network based on signalling and customer database information;
- the ability to transfer control of the call to another network element;
- interfaces to and supports management functions such as provisioning, fault, billing, etc.

Figure 10.1 illustrates the relative architectures of the softswitch and conventional circuit switch, and highlights the key differences/advantages of the softswitch.

Softswitches are meant to work in concert with application servers. Application servers will provide the interface to other more web-centred

---

[2] Source Softswitch Consortium presentation: Introducing the softswitch consortium.

applications like email, web browsers and SIP applications. It is at this point there is a slight divergence of definition of application servers. The softswitch consortium's definition encompasses media control (e.g. voice and video, through SIP) in the definition of an application server. Whilst in the IT marketplace vendors like BEA and Oracle call their products application servers. However, these products (as highlighted in Section 12.3) are purely web platforms that build on Sun Microsystems' Java-based application framework. This difference isn't a contradiction, more a specialisation. The use of an IT defined application server is possible in the softswitch environment and vice versa. In fact the softswitch consortium's application server could stand alone as a means of building hosted telephony servers, which could provide the same services as today's call centre systems (see Chapter 11).

One final note on softswitches, they are built on standard computing hardware such as Unix and Sun Microsystems' Solaris™ systems. This is an important difference when compared to the current circuit switches and a number of IN SCP platforms. The implication of this is that it should be possible to provide services based on softswitches at lower cost than the previous generation of circuit switch technologies. And it should also be possible to boost performance in the platforms by swapping out to the next processing and memory technologies, without major rework of the service software. This processing increase can be done independently of the scaling of the packet network to handle more connections or for longer duration.

10.4 THE FUTURE OF IN

We've discussed briefly the services offered by today's INs, which are implemented in the circuit switched network and briefly covered the new softswitching platforms that are coming to market. What will the next-generation network IN architecture look like? The SCP's role will be fragmented between the softswitch and application server. The SDP's role instead of diminishing could be seen as expanding, with the need to translate SIP urls into IP addresses and the need for location information to be stored. The end of what is currently referred to as IN.

The core software of an SCP, the service logic, is generally bespoke to a specific manufacturer's SLEEs; services written for one manufacturer's SCP are not readily transportable to another manufacturer's product. This has created tie-in to specific manufacturer equipment and has arguably stifled creative service development. This situation is about to change, softswitches and application servers are utilising common application frameworks (see Section 12.3) such as Java APIs for Integrated Networks (JAIN) and J2EE which allow developers to create services based on a common (manufacturer independent) set of libraries and APIs.

The Evolution of the SSP

So now we've got a feel for the change, what do the new architecture and services look like? For starters the SSP as we know it will be no more. In Part 1 (Section 5.6), the technology of media gateway control (MEGACO) and SIP were outlined. The interim next-generation SSP is a decomposed MEGACO media gateway, signalling gateway and media gateway controller (aka call server and softswitch). The longer term outlook seeing the monolithic circuit switch SSP becoming a softswitch-based architecture.

Figure 10.2 shows how the SSP will be decomposed in the Megaco architecture. The stored program controller containing the BCSM will be placed in a MGC, the peripheral connections for Time Division Multiplex (TDM) streams will be placed on media gateways and the SS7 signalling will terminate on signalling gateways. This architecture represents a stepping-stone that allows early adopters of next-generation network technology to implement existing switch functionality, but take advantage of IP trunking of voice and data internally to their own networks. From the outside world's perspective (other operators) the fabric of the early adopter's network has not changed, it still looks like a TDM switch. What is the point of mimicking a TDM switch in this way? Simply put evolution. It will take time for the current TDM-based circuit switched network to evolve into the new multimedia network of the future. In the meantime each carrier around the world will be progressing at different speeds. Some

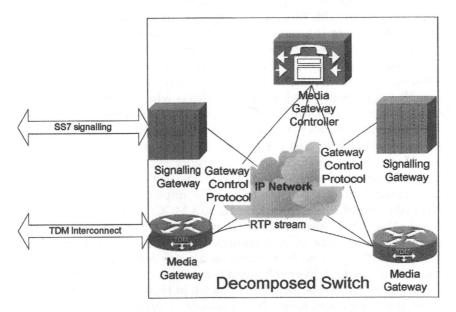

Figure 10.2   Decomposed SSP architecture

carriers will be jumping straight in with new kit, whilst others will have the financial and operational legacy of a circuit switched infrastructure that will require careful migration of customers and services, we cannot just turn the Public Switched Telephone Network (PSTN) off over night, as much as we might like to! The integrity and reliability everyone has come to expect from the current telephone network needs to be maintained, whilst its ability to cater for new enhanced services is improved.

The Megaco-SSP in Figure 10.2, as we have said, looks like a conventional intelligent network switching node (service switch point) as far as its functionality goes, if the MGC implements the same capability set and call model (BCSM) as the *old* SSP.

In a Megaco solution with the MGC implementing the BCSM of the SSP, the current SCP will be able to control the call flow as before, by utilising a SS#7 protocol stack INAP interface connected to the signalling gateway.

The signalling gateway will package up the INAP messages and use Stream Control Transfer Protocol (SCTP, see Section 5.7) for transporting the messages to the MGC. It is also possible that the vendor of current IN platforms may choose to implement an SCTP/IP interface directly on their SCP, to communicate with the Megaco-SSP. This represents an evolutionary approach to protect investment in current IN services, whilst moving the core transport to a voice over IP (VoIP) solution. This type of solution is designed to allow existing TDM network operators to slowly (and safely) migrate to VoIP architectures. New softswitches and SIP proxy/application servers can be bolted on to this architecture as and when required, to either replace or augment the capabilities of the SCP.

The advantage with Megaco for an operator already committed to a TDM infrastructure is around the trunking of the resulting call. When a number of decomposed SSPs are connected to the same IP backbone, voice connections transit from one media gateway to another, directly. In the equivalent circuit switched environment the voice call may pass through a number of switching stages. This is where the economies of an IP-based core network are realised. Figure 10.3 shows this scenario.

The resulting overall network is shown Figure 10.4. In this network it can be seen that both circuit switched and next-generation components are able to deliver services in a reasonably seamless fashion. This is achieved through the connection of the IN SCPs through signalling gateways, to what the SCP believes to be a SSP implementing a BCSM.

In reality the SCP is communicating to a MGC, which implements the BCSM. The MGC also controls a number of VoIP gateways. That completes the picture of a decomposed SSP (as examined in Figure 10.2). We will see in Section 10.5 how a voice server can be added to this architecture to deliver enhanced interactive voice services.

One additional component that is shown in Figure 10.4 is the directory server, this component forms the repository of information on how gateways can look up configuration information automatically to find out the

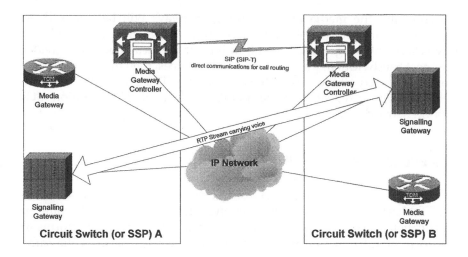

Figure 10.3 VoIP call from decomposed SSP to decomposed SSP media gateways

address of their MGC at power-up time. The media gateway control can also make use of the directory server to map, for example, message transfer part signalling point codes to media gateway voice circuit groups.

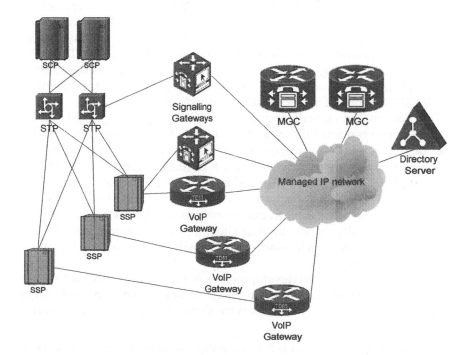

Figure 10.4 Interim network architecture

Eventually the media and signalling gateways will be removed, to make way for a fully IP-based interconnect between carriers and end devices. As the progress of the unbundling of the local loop gives greater low-cost bandwidth to premises via xDSL services, then the IP connection can be extended all the way to the end devices. This picture for the fixed network will be repeated in the mobile world as third-generation (3G) mobile networks get underway.

The Evolution of the SCP

What about the SCP? The service control point as is currently installed in the TDM network has two choices, evolve with the network to become an application service platform, or whither on the vine as new platforms in the form of softswitches and SIP application servers perform the same functions.

An interim step that may preserve the investment in IN platforms for a little while and work along the lines of evolution is to provide an interface to the IP world from the SCP to allow the SCP to invoke IP services and vice versa. It is the work of the SPIRITS (services in the PSTN/IN requesting Internet services) group in the Internet Engineering Task Force (IETF) that is looking for ways to invoke IP-based service from an SCP. The compliment to this is in the work of the PSTN/Internet Interworking (PINT) group[3] looking to call PSTN services from an IP network. The author questions the longevity or applicability of these specifications, as the examples quoted as services that could be invoked via SPIRITS can be achieved by many other means and as the next-generation services firmly fix on integrating the services on to a common IP infrastructure the need for SPIRITS and PINT services will rapidly decline.

The data in the current Service Data Points (SDPs) will have to be migrated/merged with the new databases and directories that will run the next-generation networks, either Lightweight Directory Access Protocol (LDAP) servers or enhanced Domain Name System (DNS) servers (or maybe a combination of both). This is the bleak picture for the current IN services, the new Java-based platforms will be able to provide IN style services with common frameworks (JAIN) running on standard computing hardware that allow telecoms network operators to upgrade the service performance more easily and in a more supplier independent way.

The flexibility of service development is what will arguably be the most important aspect of the move to Megaco, softswitch and application server architectures away from Service Control Point (SCP) and Service

---

[3] This group concluded their work in February 2001, resulting in RFC 2848 – SIP and SDP extensions for IP access to telephone call services.

Switch Point (SSP) implementations. The current collection of what are known as CLASS 5 features, i.e. calling party display, transfer, divert on no answer, etc. offered from local exchanges will be in some cases augmented and in others replaced by new services that offer a more integrated and flexible approach to service delivery. What are these services? My guess is:

- Enhanced voicemail. The current fixed network market for in-net answering machines is a little stale and could be enhanced by the addition of integration with email and web access. For example when you go on holiday, wouldn't it be useful to be able to log into your personal portal from your holiday resort and retrieve your messages from home, without having an international call charge; or have your messages from your mobile forwarded to your home answering service based on a set of rules that you configure through your personal portal?
- The integration of instant messaging with different devices. Right now, instant messaging is a text-based service, either through a web interface or in some instances through a short message service (SMS) gateway. Wouldn't it be a useful service to be able to integrate your voicemail and instant messaging together so that when you weren't available to accept text-based messages, text messages are deposited in your voice mailbox via text-to-speech translation, this could trigger a pager message and you could log into your voice mailbox and retrieve the message, then reply via speech recognition. Or even better, how about if someone calls your home number, but you are actually logged into a VoIP enabled PC connected to an Internet Service Provider (ISP), the call could (based on rules that allow it – for example sales calls can be filtered or calls from the boss when you're on the golf course) be redirected to the IP address you are at that instant. This type of application is an application of instant messaging and presence as described in the model for presence and instant messaging described in RFC 2778 and Common Presence and Instant Messaging (CPIM).
- Services will be able to be provisioned and changed by the customer; this will reduce cost for the network operators in the provisioning process. Sales will be possibly easier with customers browsing the 'net for new services. How will these new services be configured and tailored so easily? The use of frameworks such as J2EE and XML languages such as the Call Processing Language (CPL) (see Chapter 8) will allow easy service creation; and configuration information will be held in network directories in XML.

Closing Thoughts on IN and Mobile Evolution

Just a closing remark on the decomposed Megaco architecture, let us not forget mobile and the monumental change that is taking place to move current Global System for Mobile communications (GSM) (et al.) networks to the Universal Mobile Telecommunications Service (UMTS)/3G networks. The equivalent of the SSP in the mobile network is the Mobile Switching Centre (MSC). The MSC can be decomposed in the same way as the SSP, to give a MSC server and media gateways controlled by the MSC server using H.248/Megaco protocol. This sits easier in a mobile network that has already started to invest in packet-based technologies in the form of General Packet Radio Service (GPRS) (see Chapter 4), because key elements of the IP backbone have already received some investment in the process of rolling out GPRS services.

The alternative and long-term approach to service platforms for the next-generation network is softswitches and application servers.

The important service implication for both the fixed and the mobile worlds created by the separation of call control logic and traffic handling functions is the ability to grow these two functions independently of each other. As services get more complex and the processing requirements per call/session grow, more processing power can be installed in the soft-switch. As traffic flows demand more bandwidth per service, the transport network and number of media gateways can be increased to meet the demand. This separation of call logic also creates greater opportunity for convergence between mobile and fixed devices. The Virtual Home Environment (VHE) concepts of the Customised Application Mobile Enhanced Logic (CAMEL) specifications are carried forward into 3G networks, but by having the VHE in a server on an IP backbone, why should the only way of accessing it be from a mobile device?

Look out for location-based services as new mobile devices with Global Positioning System (GPS) capabilities slowly permeate into the market and the release of 3G services in the next 3–5 years. Expect to be bombarded with promiscuous marketing from shops and billboards that you pass on foot or in your car.

Slightly more promising for location-based services is their use when linked with other technologies such as Bluetooth and digital cordless devices (DECT). You have a single device, a telephone, Personal Digital Assistant (PDA) or other voice enabled handheld device (maybe the next-generation portable games device?). Based on location and a personalised script that runs on a network-based call server or softswitch, and the location information the device is capable of transmitting to the network, you could have, for the simple case, two access addresses (a work one and a home one). Based on your location, the time of day and any preferences such as no sales calls, the device could indicate that you are at the office and that your office access address is able to contact you, any calls to your home access address will be sent to a unified communications platform in-box. When you reach home the device indicates this and switches to the DECT transceiver and uses a fixed line out to the network. You could link your electronic diary into the script on the softswitch, so that you could have your scripts automatically inform callers that you are in a meeting and not reachable. This example service will be available within the next few years and will rely on the collaboration of fixed platforms such as evolved IN solutions and mobile platforms such as 3G service platforms implementing CAMEL services.

10.5 VOICE-BASED SERVICES

Voicemail is a valuable productivity tool and has been extensively incorporated into company Private Branch Exchanges (PBXs) and managed Centrex systems for some time. Just about every mobile phone service comes equipped with voicemail, so that even when you are out of coverage, in a meeting or generally unavailable, messages can be deposited for you to retrieve at a later time.

Call centres, the workhorse of telephone sales and customer care where agents answer calls from customers, make extensive use of voice server based services to provide customer filtering, in-queue announcements and self-serve services for customers. Call centre services are performed on specialist Interactive Voice Response platforms (IVR, also known as voice response units) either on the customer's premise or on in-net platforms provided by a telecommunication operator. The in-net platforms incorporate features of IN intelligent peripherals.

The cost of implementing the current circuit switched based services has meant that smaller customers (SMEs) have been excluded from this marketplace and only the larger call centres or companies have been able to afford to implement automated services. This started to change about 5 years ago, when companies like Dialogic started to release hardware that could be incorporated in a standard PC chassis and utilise Microsoft NT™ as the operating system to run the applications on. This change also spawned the UnPBX [MARG] or communications server. A complete phone system built out of standard PC hardware and Dialogic cards, operating as a PBX and an IVR unit.

The communication server gave the first glimmer of a world of integrated services, soon after the inception of the communications server, voicemail and email became integrated and a new service idea was born – Unified Messaging (UM). UM was meant to take the corporate world by storm as a means of allowing PC-based control of voicemail and email and allowing a new level of office automation. The initial idea was good, but it took the Internet model to bring it to its full potential in the form of Unified Communications (UC). Control of voice and email capability and a message store can now be placed on service providers' premises as a hosted service, with web and email integration rolled in.

The improvements in processing power for CPUs and Digital Signal Processors (DSPs) have meant voice recognition is becoming more viable for general-purpose use. This has enabled the creation of more usable applications, for example voice browsing where an automated voice recognition service is used to retrieve information from websites hosted off the platform. The combination of this with web integration is leading to a new set of Internet aware applications created with VoiceXML (see Chapter 8).

So after that brief introduction what are the current voice applications and how are they currently connected to the circuit switched network, and how will they change? Let's look at a simple IVR application used in call centres. There are two ways in which this can be deployed: as a Customer Premise Equipment (CPE) based service connected directly to a company's Automated Call Distributor (ACD) or PBX, or as a network-based service, connected to a carrier's class 4 (transit) exchanges.

In the case of CPE services, the IVR box is connected to the PBX or ACD via TDM trunks using either Channel Associated Signalling (CAS) or access signalling such as Q.931. CAS is quite common and additional information about the caller and access to other systems is provided by an Ethernet-based application link in the IVR platform. Applications (voice dialogues) are generally produced through supplier specific tools such as Periproducer for Periphonics platforms. These tools are drag-and-drop tools that allow dialogues to be constructed with very little technical knowledge. Additional integration to databases normally requires customer dialogue components and more technical expertise.

Whilst integration of enterprise-based voice servers are relatively simple with customer scripts specific to the call centre application. The IN-based equivalent intelligent peripheral (sometimes called a special resource platform) systems are a lot more complex, having to cope with IN integration in the form of INAP messages in addition to the call control protocol. Also a network-based approach means each platform has to carry multiple scripts (one or more for each customer), with the SCP controlling call connections and terminations.

All IVR platforms irrespective of manufacturer have the same building blocks:

- telephony processors (including digital signal processing resources for dual tone multifrequency digit detection, fax and voice recognition)
- signalling processors for interpreting the call control and INAP messages
- application processors, where the scripts that make up the dialogues execute

These components are a mixture of standard server hardware and software (most commonly Sun Sparc processors and Solaris operating system, or NT4 on PC hardware) combined with mass-produced telephony cards from manufacturers such as Dialogic and Brooktrout. In the case of the large telecommunications equipment manufacturers, custom-made cards are more widely used. These platforms are commonly placed in a VersaModule Eurocard (VME) or compact-peripheral component interconnect (PCI) chassis and mounted in 19-inch cabinets. Some larger telecoms operators have also been known to develop their own platforms.[4] The main difference between these two approaches is size. The PC-based platforms generally have fewer ports than their larger cousins based on VME or compact-PCI.

What is the future for these platforms? Out of all the services present in the current and next-generation networks, the voice applications are arguably the most used. Whether we like it or not, voicemail is a hugely successful product in the corporate environment, mobile use and also for home use in the form of in-net answering machines (see the footnote about BT). IVR for call centres is a major way of reducing agent time (and cost) and is used extensively to better target and filter callers to increase the success of routing calls to the correct agent skill (pre-call treatment; post call treatment is also common, for example reading out the Directory Number for Directory Enquiries calls). The future holds an increase in the use of these services, but also improvements in the levels of integration and service capabilities through IP. The softswitch consortium has what they call their media server framework, where media servers are controlled from application servers. Figure 10.5 depicts a possible architecture for a future voice server.

The new IP enabled IVR will be able to achieve greater port densities than its circuit switched cousins, giving service providers the facility to deliver more connections per platform than previously possible. The use of VoIP means that the new Service Providers (SPs) with no telecoms infrastructure, for example ISPs, will be able to deliver integrated voice, web and email services to customers.

The savings mentioned in the section on IN services (see Section 10.4) for call trunking will also be true for IVR services. A number of onward connect IVR services, such as calling card services currently tie up both ports on the IVR platform and valuable circuit switch ports.

The saving for IVR services is because a real-time protocol (RTP) stream can be easily redirected, whilst control of the call can remain with the softswitch, trunk costs can be reduced. When an intelligent customer interface (for example on a PC or PDA) is used, even the need for

---

[4] BT used the resources of its research labs to develop their Speech Applications Platform (SAP), this product is what the CallMinderTM service runs on. This platform was later purchased by Ericsson for future development and support.

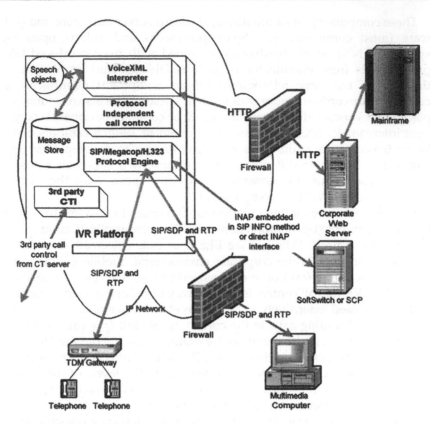

Figure 10.5 Future IP-IVR platform

follow-on call services previously having to trombone through the IVR platform for DTMF interception will be no longer necessary.[5]

The platform depicted in Figure 10.5 will be based on the hardware discussed above (PC, compact-PCI or VME), with NT/Windows 2000 for the smaller systems and Unix/Solaris for the larger systems. Current systems use a separate processor for handling the signalling requirements of the system which generally SS#7 with INAP and ISUP for controlling services. The next-generation network IVR platforms will incorporate Session Initiation Protocol (SIP) and media gateway control protocols (MGCP and Megaco protocol) for controlling the services on the platform. In Figure 10.5 this is depicted as the SIP/Megaco protocol/H.323 protocol engine. From the control perspective the platforms will also incorporate VoIP, which will be delivered over RTP. In order for the platform to monitor and control the RTP streams, Real-time Transport Control Protocol

---

[5] That's not to say DTMF interception will be eliminated – alas in-band signalling in the form of DTMF will be around for a long time to come, the POTS handset is here to stay!

(RTCP) will be used. These protocols are explained in Chapter 5. In addition to these protocols, for call centre applications, there will be a need in the interim to be able to support third-party call control. This will have to be provided by a separate module in the short term, as standardisation in this area is yet to take hold (see Chapter 11 for an explanation), eventually this will be replaced by control using SIP.

The protocols above each have their own specialities and ways of communicating that need to be made more abstract for application development. This is generally achieved through the use of an abstraction layer, depicted in Figure 10.5 as the protocol independent call control function. This has benefits as well as disadvantages, whilst it can make applications easy to develop, it can also abstract away some of the useful features a specific protocol has to offer. Because of this care must be taken when developing a protocol independent control layer.

The very top layer of the platform is where the actual voice services (dialogues) reside. Current platforms tend to use C or C++ programming languages hidden behind graphical user interfaces for constructing dialogues. This will be replaced (and has been replaced by some vendors) by VoiceXML scripts (see Chapter 8). The hope for VoiceXML is that it will enable scripts to be constructed that are independent of the platform that they are executed on. VoiceXML brings with it the opportunity to retrieve scripts and voice files from remote servers via Hypertext Transfer Protocol (HTTP). Performance considerations aside, this gives the ability for telecoms network operators and service providers to get the latest script from their customers at time of execute, this ensures the latest dialogue is always executed. To support the performance aspects of this (especially for large voice files), the next-generation network voice platform will have to contain a message store to cache the files.

The voice message store in Figure 10.5 also serves a second purpose. If the voice server is used as a component of a unified communications server, then it will be necessary to store voice messages from customers that will be retrieved later.

The final piece of the puzzle for the internals of a next-generation voice server is the ability to perform speech recognition. This will most likely be provided by extending the VoiceXML scripts to use special purpose libraries such as those provided by Nuance and their SpeechObjects™ product.

The platform will live in an IP world, and in the case of Figure 10.5, which depicts an implementation for a service provider or telecoms operator, as a hosted service. In these installations clearly security is paramount when the platform would most likely be running mission-critical voice services for corporate call centre customers. Security would have to be provided via firewalls that could be used to ensure only valid users of the service had access. Since the platform in Figure 10.5 is an IP-centric implementation, there will be, over a period of time, the need to support circuit switched connections from the existing telephony network. This will be provided either by dedicated gateways or by gateways in the service providers' network (depending on the scale of implementation).

What new services will be enabled through next-generation voice servers? A multitude is the answer and here is just a selection:

- Internet call waiting is a service currently offered as a means of doubling up your current Plain Old Telephony Service (POTS) line when surfing the net. An incoming call can be redirected to a voicemail service implemented on an IP-IVR platform, the platform could then save the message to a common in-box that could be accessed via email or the web.
- Voice browser integration for sales catalogues and websites. By utilising VoiceXML, voice recognition and text to speech, integration with websites could provide a valuable channel to online voice-based shopping. Catalogue details would be stored in XML and XML Style Language Transformations (XSLT) could be utilised to tailor the content for web browsers or voice browsing by transforming the content into either hypertext markup language or VoiceXML, respectively.
- Push-to-talk (this will be explored in Chapter 11) call filtering and self-service; when calls are generated from websites via push-to-talk buttons, the voice is carried over IP into the call centre system (some implement TDM ring back, but for the sake of discussion I'm not considering those here). Based on the customer profile (accessed as part of the customer registration process and contact patterns) an IP-IVR could be used to create a voice portal service that would be customised to the caller's needs and requirements. The voice server would be the first point of automated contact to provide in-queue announcements or even targeted advertising.
- Media conversion, one of the most important aspects of any unified communications service is its ability to cater for different access methods and thus media conversion; text-to-speech; speech recognition translation to text would be performed by an IP-IVR media server.

Figure 10.5 of the future IVR platform depicts a SIP-based signalling mechanism for session control of the IP-IVR, but what if the decomposed Service Switch Point (SSP) is still communicating to an SCP via INAP and some components of the network are still utilising circuit switched services. How could the economies of IP trunking be realised in a hybrid environment?

One solution is to allow a direct INAP protocol connection for SCP connectivity and separate VoIP trunks (gigabit Ethernet cards) (as depicted in Figure 10.5). Another method that would reduce the need

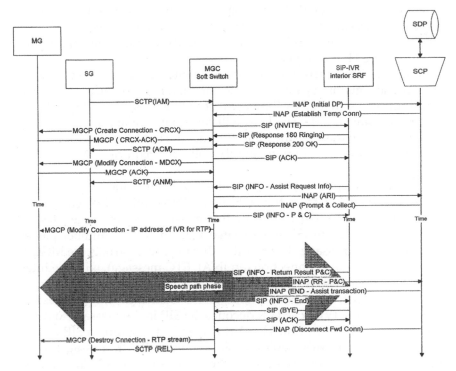

Figure 10.6 Message flows for a SIP-based IP-IVR in a MGCP environment.

for separate INAP interfaces and allow the IP-IVR to wholly reside within the IP network is to relay the INAP messages through the softswitch SSP and pass them to the IP-IVR in SIP INFO messages [RFC2976]. The IN standards allow for an intelligent peripheral to be implemented inside/ through a 'switching function'. This configuration results in the following simplified protocol exchange (Figure 10.6) based on an MGCP and SIP implementation.

The message flows depicted in Figure 10.6 show a decomposed SSP function (composed of a media gateway controller, signalling gateway and media gateway), with calls entering the IP network via a Media Gateway (MG) controlled by the Media Gateway Controller (MGC) via Media Gateway Control Protocol (MGCP). The MGC also implements Session Initiation Protocol (SIP) and communicates with the IVR via SIP messages. The SIP INFO message is used to 'tunnel' INAP messages from the SCP to the IVR. The messages have been simplified to reduce the complexity of the diagram and to aid explanation.

The IVR (labelled SIP-IVR in Figure 10.6) behaves like an intelligent peripheral which implements the IN Special Resource Function (SRF), see Chapter 3.

The SCP implements INAP over stream control transmission protocol, so can communicate directly with the MGC (without having to signal via a signalling gateway – SG), for clarity the INAP messages are shown without the SCTP envelope.

The inbound call arrives from a circuit switch connected to the media gateway and signalling gateway at the bottom (or left) of the diagram. The incoming call is signalled by the arrival of an ISDN User Part (ISUP) Initial Address Message (IAM). The IAM will contain information including the called number. In this case the MGC decodes the IAM and determines that the called number is a non-geographic number requiring IN call treatment.

The MGC suspends its processing of the call and requests a translation of the called number from the SCP via an INAP initial Detection Point (initial-DP) message. On receipt of this message, the SCP executes the Service Logic Program (SLP) associated with the called number passed in the initial-DP message. The result is that the SLP instructs the MGC to establish a temporary connection (an INAP ETC message), with the Interactive Voice Response (IVR) platform.

The MGC constructs a Session Initiation Protocol (SIP) invite message and sends it to the IVR. The IVR decodes the SIP invite message to determine which dialogue to run. The IVR responds by sending a SIP ringing message back to the MGC, quickly followed by a SIP OK message. At the same time the MGC constructs and sends a message to its MG to create a connection between the internal side of the MG and the IVR platform and associate this connection with the voice channel on the external trunk connected to the circuit switch that initiated the original call. The MG responds with an OK to indicate it has completed the request. The MGC at the same time acknowledges the SIP OK message from the IVR platform.

This has the effect of creating a context for the call between the caller and the IVR. The MG is responsible for performing the media conversion from the circuit switched world to the VoIP world.

The call can only be physically connected when the Address Complete Message (ACM) is sent from the MGC via the signalling gateway back to the circuit switch. This completes the connection in the circuit switched world. The media gateway is then instructed to create an inbound RTP stream from the IVR platform to itself, this is performed by a MGCP modify connection message from the MGC. Note at this point that the caller in the circuit switched world is receiving ringing tone from the circuit switch not from the media gateway. As of yet no speech connections are set up in the IP world from the caller to the IVR box (just the other way around). The MGC then sends a message back to the circuit switch to say the call has been answered.

On receiving the request to run a dialogue, the IVR box communicates back to the Service Control Point (SCP) that it has received the request but requires more information on what to do. This is shown as the SIP INFO message carrying an INAP Assist Request Information (ARI) message between the IVR and the MGC.

The MGC translates the SIP INFO message and sends the ARI message to the SCP. The SCP responds with a prompt & collect response INAP message. The MGC packages this up in a SIP INFO message and sends it to the IVR platform. At the same time it instructs the media gateway to send an RTP stream from itself to the IVR platform. This completes the both way connection between the caller and the IVR platform.

For the purposes of the example the caller interacts with the IVR via speech recognition and some service is requested in this way. If the caller had to interact via their touch-tone (dual tone multifrequency – DTMF) telephone keypad, the tones would be detected by the media gateway and translated into a number of MGCP messages. This would then involve the media gateway controller translating these messages into SIP INFO message to inform the IVR platform of the key presses. This would over-complicate the diagram so is not shown. Debate also takes place around the point of signalling or trying to carry DTMF messages in RTP streams.

Once the service has been decided upon, the IVR platform responds to the prompt & collect message with a Return Results (RR) message, indicating that the dialogue was completed and the result was to end the call.

The RR message is packaged in a SIP INFO message and passed to the MGC which forwards the RR INAP message on to the SCP. The SCP responds with an END message to end the ARI transaction between itself and the IVR platform. The END message is also forwarded in the same way back to the IVR. On receiving the END message, the IVR terminates the SIP session by sending a SIP BYE message to the MGC, the MGC responds with an acknowledgement to the BYE message.

The SCP finally instructs the MGC to terminate the connection to the IVR platform with a disconnect forward connection message. The MGC terminates the call by instructing the MG to destroy connection and terminate the RTP stream. The MGC then instructs the circuit switch that the call is over by sending an ISUP release message (REL).

So ends the interaction with the IVR platform.

In closing, the most exciting prospect for IP voice servers (IP-IVRs, if you will) is the fact that telecoms operators and service providers no longer have to place IVRs at a specific layer in the network (for example connected to the transit exchanges or SSPs) the trunk considerations that made this necessary go away with IP connectivity. That's not to say capacity is not a consideration any more, quite the contrary, remember the underlying issue with any voice-based communications is latency if an IP-IVR platform adds to the delay budget in any way, then voice quality will suffer. If voice quality suffers services based on voice recognition will not be viable. Research is still taking place in the manufacturers of IVR platforms to find out how to best implement speech recognition for VoIP solutions – watch this space. The longer term view is for media servers to be set up to deliver streaming video as well as the kinds of voice services explored in this chapter.

11

Call Centres

11.1 INTRODUCTION

Call centres are the place where technology advances seem to be taken up with the most gusto and where the saying 'time is money' is *so* true. A call centre environment is a pressure bottle where agents are constantly trying to serve the customer, whilst keeping their interaction times to a bare minimum, they have been labelled the sweat-shops of the 1990s and gained a bad reputation for staff morale. A call centre manager's job is to maximise customer satisfaction whilst minimising costs. The requirements for innovative technology solutions in call centres are vast and challenging. These needs have led to a number of creative solutions.

In the early 1990s Automatic Call Distribution systems (ACDs) and Private Branch Exchanges (PBXs) were the mainstay of call centres. In just 10 years we are now on the verge of network-based softACDs that integrate web, email and voice interactions with Customer Relationship Management (CRM) suites into a seamless service set. This has allowed the now re-branded 'contact centre' manager to partition the customer base according to a complex mixture of business rules and to prioritise and route inbound contacts to the best available source of help. This change has been an evolution over the 10-year period, with each year bringing incremental improvements to the technical solutions available to contact centre managers.

In this chapter we explore the rise of the call centre from a single site solution through to a multinational operation with real flexibility in the way callers can be routed, and on to the next-generation presence centre.

11.2 COMPUTER TELEPHONY INTEGRATION

Computer Telephony Integration (CTI) signalled the move from ACD or PBX only services, where call routing decisions are made by proprietary routing engines configured through text-based terminals, to a new era in call routing and configuration. The ability to utilise corporate databases such as those held by marketing to segment the inbound calls based on the market segmentation of the customer. CTI has also created the ability to integrate multiple call centres in different physical locations into a single entity. The combination of CTI and intelligent network services has created the ability for call centres to route calls across national and international boundaries transparent to the caller.

How is CTI achieved? There are essentially two types of CTI, first-party call control and third-party call control. First-party call control is where a computer takes over the role of a telephone handset's functions and links the basic call functions of a handset (make call, release call, transfer, hold, caller number presentation) to a software package that provides database integration (Figure 11.1). Examples of first-party call control are through programs such as Microsoft Outlook and contact management software such as ACT! In these examples the software controls a modem for example, with a conventional handset attached to it. Calls can be initiated from the software on the PC using point-and-click selection from the software. Or inbound calls can be intercepted and a 'screen-pop' of information about the contact displayed, based on caller number.

Third-party call control is a much more powerful capability, allowing a computer to control and monitor a large collection of telephone sets via a special interface connected directly to the PBX or ACD. This link instead of carrying telephony signalling messages carries status and control messages about all the events occurring in the PBX/ACD, from on- and off-hook messages from handsets through to call arrivals on trunks and call queue events. The control messages allow the computer system to establish new calls, terminate existing calls and even control the destination of newly arriving calls. The best way to explain third-party CTI is by

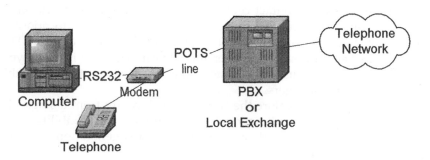

Figure 11.1 First-party call control

way of an example and a ladder diagram of the messages (Figure 11.2). First we'll explore a single site solution and then look at a more complex multiple site multi-vendor solution.

Before we rush off into an example, it's worth expanding on the issue of the 'I' in CTI. One of the biggest issues for the telecoms industry has been the use of proprietary control protocols between the PBX/ACD and CTI servers for third-party call control. Each vendor having implemented their own (in some cases more than one) protocols for third-party call control, for example Nortel have Compucall and Meridian/Symposium link on their DMS and Meridian products, respectively. Lucent have Call-Visor ASAI on their Definity product. Aspect has Application Bridge and Event Bridge on their ACD product and the list goes on.

Early work did take place on standardisation in the area of third-party call control. The International Telecommunications Union telecommunications (ITU-T) had its work on telecommunications applications for switches and computers (TASC) and American National Standards Institute (ANSI) its work on Switch Computer Application Interface (SCAI). The ITU-T and ANSI both looked at CTI as an extension of Intelligent Network (IN) standardisation and in fact the ITU-T specifications (Q.1300 et al.) were at the outset going to allow IN service invocation and were to

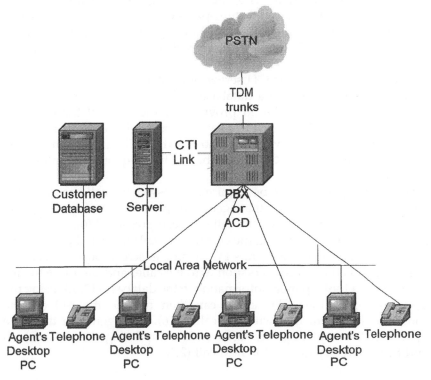

Figure 11.2 Third-party call control – single site

share the IN call model. Alas apart from some initial documentation this work floundered. The European Computer Manufacturers Association (ECMA) group have their Computer Supported Telecommunications Applications (CSTA) standards and latterly the Enterprise Computer Telephony Forum (ECTF) have worked hard to define a whole services approach to protocols, architectures and Application Programming Interfaces (APIs) for the open development of CTI services (S.200 being the CTI protocol).

Finally not to be left out both Novell and Microsoft have had a go at defining CTI protocols in the form of Telephony Server Application Programming Interface (TSAPI) and telephone application programming interface (TAPI[1]), respectively.

Alas this work has not really helped in persuading vendors of PBXs and ACDs to provide a standard set of interfaces. This has left computer telephony server vendors the job of integrating all the different CTI protocols into their products, creating a plethora of integration issues.

In the example I shall use a generalisation and simplification of the messages that flow between a CTI server and a PBX/ACD, as the paragraph above explains there are many variants of protocol, and that all perform the same role.

With reference to Figure 11.3, the following text explains the kind of interaction that takes place for third-party CTI. When a call arrives at the ACD, the ACD places the call in a holding queue that is generally determined by the dialled number (direct dial in – DDI or dialled number interception service – DNIS). This triggers a message (1) to be generated by the ACD to inform the CTI server of the call's arrival.

The CTI server initiates a database lookup in the customer database (2, 3) using the calling line id provided or the DNIS if the CLI is not present. If either DNIS or CLI is keyed to a specific customer or group of customers the CTI server then instructs the ACD to place the call in a specific queue (4). The ACD places the call in the queue and informs the CTI server (5).

When an agent of the type required to answer the call becomes available the call is connected to that agent's telephone set, and a message sent to the CTI server to indicate the call has been presented (6). This triggers the CTI server into retrieving the customer details (7, 8) and passing them to the application on the agent's PC causing a 'screen pop' to occur (9).

The agent is then in conversation with the caller and can place them on hold (10) and look up more information in the database (12, 13) using their desktop applications. The agent can then resume the call (14) and complete the interaction with the caller (16). The agent can then wrap up the call, completing any tasks such as updating the database (16) and become ready to take the next call (20).

---

[1] TAPI version 3.0 did bring H.323 and IP telephony to the PC, so all was not lost

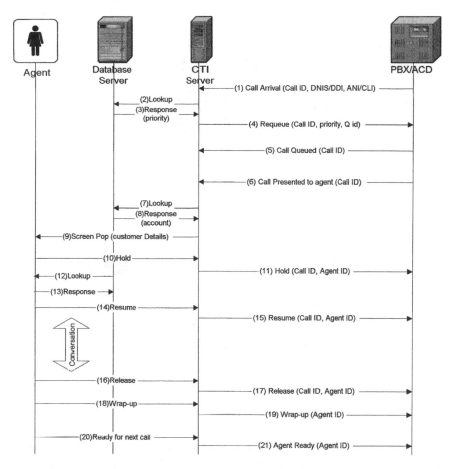

Figure 11.3 Simplified CTI message exchange

In this simple example, you can see how the CTI server is at the centre of the interaction and the ACD/PBX is constantly updating the CTI server with state changes and responding to requests from the CTI server to perform the required tasks. The agent desktop application communicates its state changes through the CTI server to ensure co-ordination of all aspects of the call. One very good reason for doing this is in the case where a call is transferred from one agent to another, the CTI server can ensure any context information is retained and passed on to the new agent.

The CTI server needs to retain call state information and like the IN needs to have a call state model – in the case of IN this as we have discovered earlier in the book is called the Basic Call State Model (BCSM). In the case of the CTI server the call states have to be compatible with the particular ACD/PBX call states to ensure step-by-step tracking of call flow.

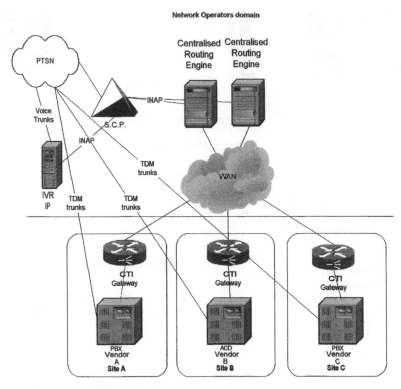

Figure 11.4 Multi-vendor, multi-site CTI architecture

In a multi-site multi-vendor situation things get a little more complex, the CTI server needs to maintain state for multiple agents spread across multiple ACDs each with a different call model and set of CTI messages (Figure 11.4). In the two most prominent products on the market, a gateway device performs the mapping of call state and proprietary third-party call control messages. The gateway device translates the vendor-specific messages into an internal generic message set. The internalised messages are used to update a centralised control point, which contains the queuing and routing rules to ensure multi-vendor and multi-site co-ordination. This type of solution works reasonably well for an enterprise situation, however, calls arriving at a specific location that are then onward routed to an agent on another ACD/PBX in the group, will have to be connected between the two ACDs (hair pined or tromboned). This onward connecting of calls costs money in the form of dedicated interconnecting trunks.

The solution to the problem of dedicated trunks is to utilise the network operator's IN to perform pre-routing decisions. More efficient call routing can be achieved this way, allowing calls to transit the public network before reaching the PBX/ACD that the agent destined to handle the call resides on. In this type of solution the network operator also commonly provides 'in-net' Interactive Voice Response (IVR) services so that calls can be more efficiently pre-routed. Two types of IVR control are common: the IVR can be configured as an IN Intelligent Peripheral (IP) or third-party call control (CTI) interfaces are used to communicate with the scripts in the IVR platform. An example of the architecture is shown in Figure 11.4.

11.3 THE FUTURE FOR CTI

CTI is undergoing a significant boost from the integration of voice and data on a next-generation network. The first generation of CTI products brought about the control of circuit switch based ACD and PBXs and allowed for increasing complexity over the control of call routing through the use of information from customer databases. Second generation CTI allows for the integration of systems that provided a view of the customer and their relationship with the organisation (labelled Customer Relationship Management (CRM) tools), web collaboration and email routing with conventional Time Division Multiplex (TDM) platforms. Next-generation CTI is combining Voice over IP (VoIP), electronic CRM (eCRM) and IP-voice servers (see Chapter 10) to bring a level of integration beyond that which was previously possible to create packet telephony call centres.

The next-generation CTI servers will combine the ACD routing capabilities, but go beyond this to provide a truly integrated solution for 'contact' routing. Presence is a new service that has gained significant interest in the telecommunications and Internet industries as a means of combining location-based information with the instant communications of Instant Messaging (IM). If we consider the concept of what a contact centre embodies, it is in fact a presence service [DYNA]. The desire to communicate via some means (Personal Digital Assistant (PDA), VoIP, email, text-chat, etc.) combined with the current state of the communicating entities. The work on standardising presence in the Internet Engineering Task Force (IETF) is at the time of writing in its early stages and a number of proposals are being discussed, one of which is the use of session initiation protocol (SIP) to support presence.[2]

The use of SIP in contact centres is in the author's opinion the nirvana of CTI and utilising SIP as a protocol to support presence service creates the idea of a 'presence centre'. The presence centre goes beyond the contact centre ideas and capabilities and provides customers wishing to contact an organisation with much more information about the status of the presence centre agents they're trying to reach. SIP's inherent scalability

---

[2] IETF working group called SIMPLE.

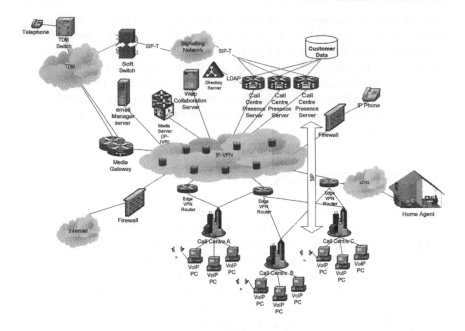

Figure 11.5 Presence centre architecture.

also makes for a very happy vendor community, where servers can be made to scale from small 20-seat operations to meet the demands of the Small to Medium Enterprise (SME) market, to the tens of thousands of seats capable systems to meet the needs of the xSP marketplace.

The architecture for a presence centre is shown in Figure 11.5, this is a not a TDM solution, there are no circuit switching components to the softACD application server (presence server). All communications between caller, agent and softACD are performed by SIP messaging. In addition the web servers utilise a Java 2 Enterprise Edition (J2EE) framework with SIP servlets to allow push-to-talk functionality, where a button on a web page can be used to initiate a conversation between the person browsing the website and an agent, either as voice communications or text-chat, with SIP being used to establish and terminate the session(s).

eCRM applications are hosted on the application server, with access to enterprise databases and Enterprise Resource Planning (ERP) systems. The application servers that will make this possible are already being developed in the form of products from Oracle, such as their 9iAS™ application server product and BEA systems' WebLogic™ server.

Home agents are easily accommodated through the use of Digital Subscriber Line (DSL) and Voice over DSL (VoDSL) services. Where appropriate a multi-vendor environment (for example SIP phones from different vendors establishing connections via the presence server) can also be accommodated via the open nature of SIP.

To explore this architecture further, let us take each of the major components in turn. Call centres containing agents will be able to exist in locations that are remote from the network where the servers that run the applications that perform the contact routing reside. The call centres will not need large amounts of infrastructure aside from desks and multimedia equipped personal computers.

The applications that the agents (or customer service representatives) require to perform their role will be able to reside within the network, to make modification and role out of new features easier. This also opens up the opportunity for IT managers to outsource the support of this to applications, to Application Service Providers (ASPs).

All voice connections between the agents and customer will be transported through an Internet Protocol (IP) network that can give Quality of Service (QoS) guarantees. This will be made possible through the use of Multi Protocol Label Switching (MPLS) and high capacity fibre links. It will be possible for service providers to provide secure service to multiple customers from the same infrastructure in this way too; MPLS is able to segregate the traffic for each customer to create a Virtual Private Network (VPN), which is the cloud in the centre of Figure 11.5. Multiple remote call centres will be able to function as one large virtual call centre in this way.

In Chapter 10 the extension of IVR platforms to support integration into the IP infrastructure was explored. Through these new IP-IVR media servers, it will be possible for service providers to run contact filtering services using speech recognition to assist customers with their enquiries, even fulfilling the enquiry without having to speak to an agent. Whilst existing IVR platforms provide these facilities IP-IVR will bring economies to the connection of customers to the IVR and agents. IP-IVR through the use of VoiceXML will also bring a closer integration with websites, allowing customers to get access to the same (consistent) information irrespective of how they chose to contact an organisation.

Contact centre presence servers (or software ACDs) are the intelligence that will make all this possible. Software ACDs will provide SIP-based media control that will connect customers with IVR systems and agents (using presence state updates to keep track of the status of all the components it controls – see Chapter 12 for more on presence). They will also communicate with softswitches to enhance call routing from callers outside the IP network. SoftACDs will be able to route telephone calls, web-based text-chat requests and emails. They will, by virtue of technologies such as the Java Connector Architecture (JCA) and Java database application programming interface (JDBC), have the ability to integrate with a large variety of Enterprise Information Systems (EIS) (see Chapter 12 for more details on the feature of the J2EE which enables this). The integration with EIS systems will enable the rules used to route contact requests to take into account real-time customer information.

Web integration will be provided by application servers (see Chapter

10) that provide web collaboration features, allow customer and agents to simultaneously view the same web pages and converse in real-time about the products on offer.

And finally home agents will be able to use the same facilities as their colleagues in the contact centre, since all the applications are hosted in the network on application servers. DSL will provide the bandwidth necessary to seamlessly carry voice and data to home agents.

In summary computer telephony integration is blazing the trail of convergence towards the next-generation network services, so much so the US *Computer Telephony* magazine has re-branded (June 2001) to *Communications Convergence* magazine. It could be argued that CTI forged the path to convergence even before the ideas of a next-generation network were formed. There is no doubt in the author's mind that contact centres will continue to push the boundaries of computer telephony integration and will be the standard bearer for convergence, with the release of convergent eCRM and telephony server platforms, call centres are the future battleground for the next-generation network service provider.

12

Internet-Based Services

12.1 INTRODUCTION – THE MOVE TO HOSTED SERVICES

It all started with Internet Service Providers (ISPs) offering to host a company's website and email. ISPs can offer economies of scale and could cost effectively maintain a permanent connection to the Internet and offer dial-up solutions to their customers. The customers in return get a permanent presence on the Internet for their website and permanent email delivery service.

In the corporate environments the move from centralised server and mainframe environments to peer-to-peer networks with local workgroup servers had started the return to centrally managed servers. In the early to mid-1990s, technologies such as the Citrix WinFrame™ products enabled resurgence in the centralised management and delivery of enterprise applications.

These two technology solutions combined with organisational desires to outsource functions of business activity, seen to be non-core, led to the opportunity for application hosting services. The idea of thin client computing was born and the web browser became the ultimate thin client application. Architectures such as the now famous three-tier architecture of client, web server and back office database engine spawned the use of server side scripting and Java applications running either on the web server or in the web browser (applets in this case) and the n-tier architecture model was born. This model has given birth to application servers and the architecture of call servers and softswitches.

This evolution of technology has enabled the growth in the service provider marketplace (xSPs as they have become known) including:

- ISP – Internet service provider.

- ASP – Application Service Provider.
- WASP – Wireless Application Service Provider.
- WISP – Wireless ISP.
- AIP – Application Infrastructure Provider.
- ITSP – Internet Telephony Service Provider.
- CSP – Content Service Provider.
- Telecom hotels – where web, email and telephony services are provided with a building as part of the fabric like other utilities such as water and electricity. These hosting centres are sometimes shared, for example in the UK, London Docklands is host to Telehouse, a building providing shared services for a number of ISPs with international interconnections to other Internet hosting centres and tier 1 ISPs (those ISPs with a direct connection to the main Internet backbone or those ISPs who actually own some of the main Internet backbone that other ISPs buy capacity on).

The forecasts for the growth and development of this market area were extremely bullish in 2000 with all the major research organisations forecasting huge potential revenues from this market opportunity. Whilst the figures may well be exaggerated the use of these kinds of services will increase as confidence in both the technology and the ability of the SPs to deliver on quality and security increases and of course the ability to bill for the usage of such services. The model itself will continue to provide a valuable architectural concept for the delivery of next-generation network services, as it encompasses the desire for the mixture of processing power both in the network and in intelligent devices.

All xSPs share the same characteristics, those of: offering services over a network connection, managing the services on behalf of the customer, providing the same service to many customers to gain economies of scale and generally levy a recurring fee (monthly, quarterly or annually).

The concept on a rental-based service delivery is not new and facility management companies have been providing similar services for some time. The newness is in the delivery mechanism – over a network. Telecommunication service providers have been providing this service for some time and in some countries the telephone line and handset were rented from the telecoms operator. This arguably puts telecoms operators in a strong position as their internal processes and business systems are set up to manage a rental approach to service delivery.

The key message from the previous paragraph is that the telecommunications network operators who have decided to move their infrastructures to deliver next-generation network services are moving to a very strong position in the marketplace. Not only will they be able to offer more services such as application hosting (ASP), but they will be able to offer infrastructure provision in the form of virtual private networks and dedicated servers in hosting centres. This means telecommunication network operators moving in this direction stand to have the capability to offer the service providers services. In Chapter 14 the idea of a virtual service provider is explored and other concepts of how the market may develop.

This very brief description outlines the change from the basic web server, to a new application server and content delivery framework. This framework will sit alongside and even underpin the call server frameworks to enable the combination and collaboration of telephony services and enterprise applications in a network-based hosted environment. Three components will form the basis of all new services, two of which we have explored: softswitches and application servers. The third component is infrastructure. Session Initiation Protocol (SIP), Extensible Markup Language (XML) and Java will be the dominant technologies that will bond this new service orientated environment together.

12.2 PRESENCE

In Chapter 11 call centre services were explored and their evolution to presence centres was postulated. The following section explores the concept of presence, and by the end the reader should be able to understand how presence as a concept and a service could underpin many services clambering for the throne of the elusive 'killer application'.

One of the most compelling services to come from the Internet recently has been Instant Messaging (IM). All the major portals and ISPs (MSN, Yahoo, AOL/Compuserve, LineOne, etc.) offer an IM service and client application. The early releases of IM clients were incompatible in a bid to create tie-in to a particular portal (not unlike the browser wars that ruled the 'net a little while back), however, more recently a number of clients have started to appear that are compatible with the others.

The Internet Engineering Task Force (IETF) have got involved and at the time of writing two informational RFCs have been written and three Internet drafts under the Instant Messaging and Presence Protocol Working Group (IMPP-WG). The working group's aim is to define protocols and data formats for the IM and presence community, to allow robust Internet-scale IM and presence applications to be constructed. All was not easy with the IETF during the formation of this group and a number of different parties all proposed different approaches to IM and presence solutions, this finally resolved, work is still underway to complete the standardisation effort.

The informational RFC 2778 describes presence and IM systems as systems that allow users to subscribe to changes in each other's state and for users to be able to send each other instant messages. In the case of current IM applications this state is normally represented as some form of 'buddy list' and instant messages can be sent to and from both web clients and Short Message Service (SMS) or Wireless Application Protocol

(WAP) interfaces. IM clients also currently integrate access to email and an online diary/calendar.

So what is the difference between IM and presence? Current IM applications only really reflect whether you are online or offline. Presence services will reflect more dynamic behaviour linked to location-based services and with information about the type of terminal you are using, the type of communications you are capable of receiving.

This means the integration of technologies such as: voice over IP, IM, mobile handsets and Personal Digital Assistants (PDAs), Public Switched Telephone Network (PSTN) services, email and perhaps even online games. One of the early examples of this kind of presence client is from Hotsip, their Active Addressbook™ client. It allows you to instigate multi-player network games whilst communicating with friends either via text-based chat session or voice/video over IP. It is designed around the SIP protocol.

What are the key components to a presence service? Figure 12.1 shows the key components of a presence service (this is based on RFC 2778).

The presentity software is responsible for informing the presence

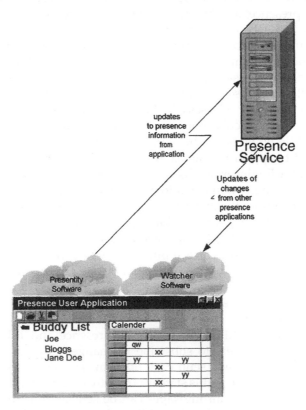

Figure 12.1 Presence service components

service of the changes in the status of the user, called the 'presentity'. In the case of a user application as shown in the figure this is reflecting the changes the user instigates by selecting status changes in their Graphical User Interface (GUI). However, in an embedded application say in a mobile device, this could reflect location updates, availability of different communications medium (say I just walked into a room that has a wireless Local Area Network (LAN) or a Bluetooth LAN) or user changes.

The watcher software is responsible for receiving updates about 'buddies' and for instigating the subscription to the buddy's information with the presence service. The watcher can also do one-shot updates by polling the presence service for an update. The watcher software is known as either a 'subscriber' or a 'fetcher' for each of the operations outlined in the previous sentence.

Clearly for this model to work and information to get exchanged between presentities and presence services, a protocol and the structure of the messages to be exchanged must be defined. These topics are still items for discussion in the IETF-WG, but SIP is one of the strong proposals for the protocol and a draft proposal has been made available for the message formats for IM and presence (draft-ietf-impp-cpim-nsgfmt-03.txt).

We started with a bold statement on presence, that it will be a key service enabler in the future next-generation network. Maybe after the previous description readers can form their own view as to the future of presence?

12.3 APPLICATION FRAMEWORKS

Introduction

Let us be clear what is meant by application frameworks, as some clarity is required. Object-oriented application frameworks are a software engineering technique for presenting a skeleton of object classes that can be used to construct a multitude of applications all based on that common underlying framework. For example Sun Microsystems have the Swing Java classes for constructing GUI applications that will 'look and feel' the same whether the application runs on a Mac, a PC or under the X-Windows system.

Much larger frameworks are starting to appear for the development of a whole group of enterprise applications, examples of these are Microsoft's .NET (pronounced dot-NET), Sun's Java 2 Enterprise Edition (J2EE™) and the Java API for Integrated Networks (JAIN), an incentive to bring an open framework for advanced telecoms service development as part of the Sun Java Community Process.

It is these larger frameworks that are extremely important in the development of next-generation network services. By providing a common framework for a service and application development, it opens the telecommunications networks of the future up to a wealth of developers and freedom. We have discussed the fact that whilst Intelligent Networks (INs) improved on the stored program controller concept of a switch, the applications written for a specific supplier's Service Control Point (SCP) are tied to that SCP. Developers have to be skilled in that narrow area of the supplier-specific implementation of the IN Service Logic Execution Environment (SLEE). Incentives like JAIN allow third parties to get in on developing services for telecommunications networks, with transportable supplier independent skills, this will greatly reduce the cost of development and it is hoped nurture innovation.

.NET is a framework for creating componentised distributed applications that can communicate both between components and between whole application instances across a wide area network. Its relevance is yet to be tested by the market, but Microsoft is looking to the .NET framework to provide application service providers with a mechanism for developing applications that can be hosted and distributed easily. J2EE provides similar ideals and builds on the already strong concepts and wealth of existing Java Application Programming Interfaces (APIs) to deliver a framework for distributed web-based *n*-tier applications. Oracle for example has put a lot of effort into extending its database technologies with the J2EE framework in their Oracle 9i application server product.

This area is still very young and a lot of work will need to be done to ensure the opportunity is realised in real-world valuable applications and services. That said the industry is backing the frameworks approach and a number of key telecoms manufacturers (Nortel, Nokia, Alcatel, etc.) are all committed to the Java incentives. It is clear that .NET will find its way into the corporate desktop, if ASP style hosting takes off.[1]

Java API for Integrated Networks (JAIN)

The marketing information on JAIN claims the initiative brings service portability, convergence and secure network access to telecoms networks. To address each in turn, we have already exposed the flaw in IN services, that of non-portable applications. JAIN overcomes the service portability barrier through the Java write-once, run-anywhere philosophy that Java

---

[1] Some might say hosting services are already a reality, the author is not sure if this is really true or just more hyperbole, whilst the web hosting business is significant the application hosting business is still in its infancy. Small to Medium Enterprises (SMEs) will clearly benefit from outsourcing their applications support to reduce IT spend and using the services of an xSP has significant benefit for a distributed company.

realises through its use of an interim code level called byte code, which runs in a Java byte code interpreter and by providing a 'standard' common API framework. JAIN delivers convergence through the implementation of multiple protocol stacks (TCAP, ISUP, MAP, INAP, MGCP, SIP, SIP sevlets, MEGACO/H.248 and H.323), allowing through the JAIN framework the implementation of softswitch functionality on any platform that has the Java runtime engine installed. The Java sandbox is used as a security mechanism for controlling rogue programs. By constraining applications to a set of functions they can perform and provide secure remote method invocation procedures, JAIN provides a secure environment for network applications.

The JAIN work is split into two sets of APIs: application APIs and protocol APIs. We've already covered the protocols, the application APIs are: JAIN call control (JCC), JAIN Co-ordinations and Transactions (JCAT), a JAIN SLEE (cf. the IN SLEE), a JAIN service provider API for trust and security management (SPA), a SPA mobility API, a SPA presence and availability API for presence and IM, and a JAIN service creation environment. Some of these APIs are at the time of writing still under development. Figure 12.2 gives an overview of the relationship of these components.

JAIN really builds on the tried and trusted (A)IN model for external service construction and control, but with the already extensive Java API library to call on, to be able to extend and create integrated applications in an 'open' SCP. JAIN is not the only incentive. The third-generation (3G) mobile world has been working on an Open Service Access (OSA) framework definition which allows third-party applications to be 'plugged' into the Universal Mobile Telecommunications Service (UMTS) mobile networks. Both the JAIN work and the OSA work have been influenced by the Parlay specifications. In fact the JAIN service provider APIs are a Java implementation of the Parlay APIs.

The Parlay group is an industry consortium formed in 1998 by BT, Microsoft, Nortel Networks, Siemens, and Ulticom (formerly DGM&S Telecom). The aim of the group was to create an open framework of APIs to support the development of communications applications. Since the inception of the group, a number of other companies such as Cisco, AT&T, Microsoft and IBM joined.

The aim of the Parlay APIs is to provide a network independent application development environment by the specification of programming language independent APIs, so you can see why JAIN and OSA have looked to them for inspiration.

The Parlay specifications cover the following areas:

- call processing
- connectivity management
- framework, this covers aspects of security and management
- messaging
- mobility

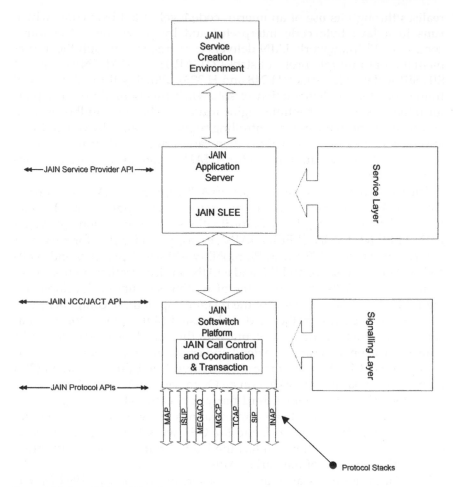

Figure 12.2   JAIN architecture

The call processing APIs specify a generic call control service, which provides a third-party call control model. That in theory could provide a framework for developing a call control programming library for the major protocols present in the current and next-generation networks (INAP, ISUP, H.323, SIP, etc.). There are some camps that argue this approach is flawed because no single specification can cover all the specific strengths of a particular protocol and SIP CGI presents an alternate approach that tries to address this concern (see Section 12.3).

Java 2 Enterprise Edition (J2EE)

J2EE does for web servers what JAIN does for SCPs. It is a collection of APIs bundled into a framework that sits behind a web server and allows more feature-rich enterprise applications to be built on the web model. J2EE provides tools and APIs for both server side and client side development. The J2EE model defines a set of tiers, a client tier, a middle tier and an enterprise information system tier (see Figure 12.3). Each tier contains a set of Java capabilities either frameworks such as the enterprise Java beans (EJB) container, sevlets and the web container or programming capabilities such as Java server pages connected to web pages via Hypertext Markup Language (HTML) tags or as embedded code for presentation of the information from the back office systems.

Servlets are web server extensions to a web server that allow Java programs (implementing the servlet API) to be loaded dynamically into a Java runtime environment of a web server, replacing Common Gateway Interface (CGI) scripts.

Integration to back office systems is via the Java Database Connectivity (JDBC) APIs for connecting to relational databases and the new specified Java connector architecture for connecting to Enterprise Information Systems (EIS). EIS are as Figure 12.3 shows, systems such as enterprise resource planning systems, customer relationship management systems, business support systems, etc.

J2EE also provides programming libraries for access name services such as Domain Name System (DNS) and Lightweight Directory Access Protocol (LDAP) directories (see Chapter 9), and access to email systems via the JavaMail program library.

J2EE also provides APIs for transaction management, Remote Method Invocation (RMI) and message services. The transaction management API provides Java classes to help in the construction of applications that need to co-ordinate a number of distributed events that result in an atomic action being taken, maybe writing a customer record to a database. The

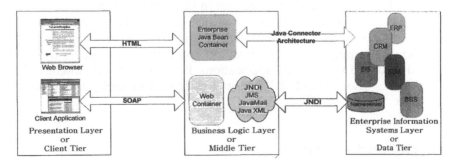

Figure 12.3 J2EE framework tiers

RMI API allows synchronous communication between distributed applications to be programmed with very little knowledge of distributed applications. It is an enabler for building *n*-tiered distributed applications. The message service APIs provide similar service to the RMI libraries, except message-based services are asynchronous. A message can be posted to a queue and forgotten about. The application posting the message can do other things.

J2EE is a complete application framework for the development of multi-tiered enterprise applications.

This is a very short description of the kind of services J2EE provides. However, this brevity belittles the real power of J2EE and its significance for the development of hosted services in next-generation networks. Java's strength of write-once, run-anywhere creates a fantastic opportunity for third parties to develop applications that will run on a variety of vendors' platforms, increasing the opportunity for diversity and richness of new services.

.NET

Microsoft's .NET is an idea designed and built around XML. Application communication is provided via Simple Object Access Protocol (SOAP). The SOAP specification whilst originated in Microsoft is now a World Wide Web consortium (W3C) standard and is defined as an XML document specification. The W3C specification means that .NET applications are not restricted to Microsoft only platforms, for example a SOAP server implementation already exists for Linux. Microsoft is concentrating its efforts around their BizTalk™ server and their HailStorm concept of web services. This is likely to be a major factor in the distribution and aggregation of new services.

The .NET framework is clearly Microsoft's answer to J2EE, as the .NET framework also comes with its equivalent of the Java runtime called the Common Language Runtime (CLR). Code is compiled into Microsoft Intermediate Language (MSIL) to run in this environment.

At the time of writing .NET was just getting started and looks to promise an alternative to the J2EE model for development and will provide a stepping-stone for Microsoft application developers to move into distributed web-centric applications. For those who are interested in finding out more about .NET then I suggest you look at Microsoft's website.

SIP CGI and SIP servlets

The Common Gateway Interface (CGI) for Session Initiation Protocol (SIP) is not really a framework more an extension of the mechanism used by web servers to invoke server side scripts. Made popular by Dynamicsoft, it is a mechanism that can be easily built into an application server/proxy server/registration server platform that can allow users to upload scripts written in CPL (see Chapter 8), that execute on the server.

The CGI used on web servers evolved as a mechanism for generating dynamic content and processing form submissions from web pages.

SIP CGI works like a traditional web (hypertext transfer protocol) CGI, when a request (in this case a SIP INVITE message) arrives at the server, the message and all its parameters are passed to a script through the standard input and sets a number of environment variables that contain information about the message and where it came from. The script on the server executes and uses the information stored and the standard input to make decisions on its behaviour.

In a HTTP CGI implementation the script generates a response (which in most cases is HTML to send back to the client that created the original request) and exits. SIP CGI is different in this respect. The output of the script can actually be an instruction to create a new request. In the case of a SIP proxy server (see Chapter 5) this is quite likely to be the result of the execution of the script.

An HTTP CGI script is normally transaction driven, with each request a single response is made, with no context as to any previous transactions. With SIP CGI however this would not be useful, since the SIP CGI script can generate a request itself, then sometime later a response would be expected, clearly the original script that generated the request needs to be informed of the response. State is maintained through the use of 'script cookies'. The cookie is passed as part of the response to the CGI server in environment data, this cookie is passed back to the script when another request arrives. Armed with a cookie and the new request the script can maintain state between transactions [RFC3050].

Since its creation the web CGI has been augmented with servlets and server side scripting. Servlets (see Section 12.3 on J2EE for a description of Java servlets) are a Java replacement for the CGI, instead of the web server running a separate process to execute CGI code, the Java virtual machine on the web server runs a servlet code.

SIP servlets extend the idea of Java servlets to SIP messages, allowing applications to quickly and easily have access to all the information in the SIP messages. This, it is hoped will speed the delivery of server side SIP aware applications. The SIP servlet API is now part of the Java community programme and the hope is that SIP servlets will form part of the Java 2 Standard Edition and Java 2 Enterprise Edition APIs in the future. This small addition to J2EE (if it is incorporated) would make it possible for all application servers, that are J2EE compliant, to be capable of becoming media control/call servers.

OSS-J

There is one last word on frameworks that is possibly even more important than what we have already covered. It is the OSS through Java initiative. This initiative is looking to simplify the complexity of Operational Support Systems (OSS) through the use of J2EE and specifically the enterprise JavaBeans™. The OSS-J community programme is looking to build on work already underway in the Third-Generation Partnership Programme (3GPP) and TeleManagement Forum and to provide a set of Java APIs, source code and component models that will be freely distributed. Conventional OSS and Business Support Systems (BSS) involve lots of integration issues between different vendors' equipment; OSS-J looks to address these issues. If it is successful it may signal a new era for management of complex services and just what is needed and necessary to provide the important *glue* between the new services and the operation, maintenance and billing for them.

The OSS-J initiative is currently focused on deliverables for the 3G mobile networks and specifically the areas of:

- service activation (create, amend and cancel)
- quality of service (support, operability, serviceability and security) and
- trouble ticketing (fault tracking and resolution)

The Java community process demands the following deliverables: APIs for each of the areas above, a reference implementation and a technology compatibility kit (essentially a toolset with a collection of tests to prove other platforms conform to the specification). More detail can be obtained from http://java.sun.com/products/oss, and I suggest the interested reader check here for all the latest developments.

In closing, the frameworks presented in this chapter will form the basis for many of the xSP application server platforms and softswitch services that were discussed in the previous chapter. XML is probably the most important component incorporated into all the frameworks discussed, it will form the glue that describes the services, their configuration and the protocols they use to communicate. The big names are keen to move in this space and people like Oracle, Dynamicsoft and BEA (and others) already have key products in the application server marketplace. The link between application servers and softswitches is the next step that will complete the next-generation network architecture.

13

Bringing it all Together – the New Network Architecture

13.1 INTRODUCTION

In Part 2 of the book we have explored some of the key existing services and how they may progress. In this chapter I would like to take you a little way into the future, for a view of what the network may look like and explore how a service or group of services may fit together to form a useful example of where next-generation networks will take us.

Let us go over the components of the next-generation network:

- Media gateway and signalling gateways, these are the interfaces back to the Time Division Multiplex (TDM) world of the current Public Switched Telephone Network (PSTN) and Public Land Mobile Network (PLMN).
- Media gateway controllers (call servers), softswitches and application servers. These are where the new services execute.
- Media servers (or Internet Protocol Interactive Voice Response (IP-IVR)), these are the components that provide the voice of the network and automate a number of voice enabled applications.
- Directories, these will store all the configuration information about services, users and equipment inventory, for example:

  - Application-specific configurations for each client/customer.
  - Personalisation information on a per user basis (buddy lists, preferences, usage statistics, favourites and bookmarks).
  - Device information and capabilities.

- Location of application components and whole service instances.
- Service translations (freephone style service translation info).
- Security tokens (access control lists and encryption keys).
- User location information.
- Virtual Home Environments (VHEs).
- Service tags, flags to indicate what the user is allowed to do. For example if someone hasn't paid their subscription fee, then a flag could be present to prevent them from using any billable services.
- User scripts (for example written in Call Processing Language – CPL).
- Application scripts (VoiceXML scripts say).
- Agent skill sets for call centre applications, the profile of an agent's capabilities could be held.
- Call routing rules.
- Billing information and usage records.

- The directory will hold the information in Extensible Markup Language (XML), because XML brings with it the power to represent the data in a structured way that allows the display and transformation of that data to and from any device.
- Packet network, the infrastructure that supports all the components above. It will be using protocols such as Multi Protocol Label Switching (MPLS) to both segregate traffic flows and to manage quality of service aspects.
- Transport infrastructure, this will be fibre optic cables providing high amounts of bandwidth managed via Dense Wave Division Multiplexing (DWDM) and broadband copper for low cost Small to Medium Enterprise (SME) and residential access. For mobile devices the access will be via Universal Mobile Telecommunications Service (UMTS) data and voice services.

Now that we have the components how do they fit together? They say a picture speaks a thousands words, so I'll use a picture to give some idea of the architecture.

13.2   THE NEXT-GENERATION NETWORK ARCHITECTURE

In the introduction to this chapter we highlighted the components that will make up the next-generation network architecture. Figure 13.1 shows a view of the components and their relationship to each other to form the physical architecture of the network.

The technologies such as DWDM and MPLS will form the basic infrastructure (glue) that will interconnect all the nodes. MPLS will be used to segregate the different type of data that will be transported on the infra-

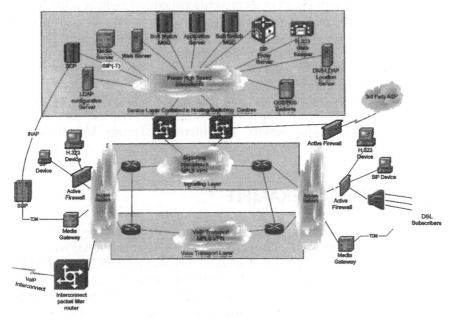

Figure 13.1 The next-generation network architecture

structure. The data will be one of two types: media streams (voice, video and other forms of content) and control data (signalling).

The bottom-most layer will be the media transport layer, in the case of a next-generation network to replace the circuit switched network of telecoms operator, this layer will carry lots of telephone calls. This is shown in Figure 13.1 as the voice transport layer. Part of this layer's function will be to groom incoming traffic from the broadband local loop, for example from cable and digital subscriber lines. The media traffic from these customers will be delivered over the media layer and their signalling messages from the intelligent access devices, passed to the signalling layer.

Above the media transport layer will be the signalling layer, carrying all the control messages to and from media gateways and their media gateway controllers, encapsulated Signalling System number 7 (SS#7) signalling messages carried from signalling gateways to media gateway controllers and softswitches and session initiation protocol (SIP) messages between SIP clients and proxy, redirect and SIP application servers.

The very top layer will be where all the servers reside to allow fast communication between distributed applications and between different service platforms. This layer will also carry all the information to the operational support systems and business support systems for maintenance and billing. This layer will also contain all the directories and databases that will store all the location and configuration information about customers/users and the services. This layer could also hold the

Service Control Point (SCP) of a conventional circuit switched Intelligent Network (IN). The SCP in this case would have implemented an Internet Protocol (IP) interface and would no doubt be communicating (via stream control transmission protocol) with a media gateway controller or soft-switch masquerading as a service switch point. This would provide old and next-generation network service integration.

The final and probably the most important parts of the network will be the firewalls and routers that interconnect the networks of different network operators and service providers. In Figure 13.1 these are shown as firewalls and the interconnect packet filter router.

13.3   A SERVICE EXAMPLE

Imagine a couple of young professionals in a few years' time going about their hectic lives, both have Personal Digital Assistants (PDAs) with voice (UMTS and Digital European Cordless Telephony (DECT)) and video capabilities, plus close proximity Bluetooth capabilities. The service provider they got their PDAs from provided them with basic packages containing a bundle of the following services: location-based advertising from a selection of reputable high street names, real-time stock, traffic and news feeds, access to an online diary/agenda and presence service, with a bundle from their UMTS and fixed line providers that allows them to use their PDAs as fixed line handsets in their homes, because they get free local calls in the evenings from the fixed line provider and of course the now ubiquitous Unified Communications (UC) package.

Peter and Jane are both in work, but it's getting close to the end of the day and Peter feels like he's had a tough day and would like to eat out and see a movie. He checks Jane's online diary/agenda to see if she has any late meetings (Jane has previously allowed Peter read access to her diary/agenda). She doesn't, so Peter looks up on the web what the local cinema is showing. There is a couple of showings at the 9:30 p.m. slot for the latest summer blockbuster, which he thinks Jane might just be persuaded to watch. So he puts a one hour reservation on two tickets, which he will need to confirm.

He's not sure what food he would like, so he decides to have a stroll into town to have a look around the restaurants to see what takes his eye. On leaving his office building the network detects that he has left his office cell, and this triggers a script that diverts all calls to his office number directly to his UC service. The service script is smart enough (its in the network, so can interrogate the source of contacts) to distinguish work calls from personal calls. Peter has set up an additional list that allows calls he has nominated as personal calls that arrive at his office number, to be routed to his PDA and not the UC platform.

Passing the Italian restaurant his PDA flashes up the specials for the day, this catches his attention and he really ought to find out if Jane would like Italian and the movie. He goes to his presence service where he has Jane in his buddy list. He sees that Jane is still at work and set to be reached by text messages only. Peter decides Jane must have someone with her who caught her at her desk before she set off for home.

Peter sends Jane an instant text message with the film details and the Italian restaurant menu, plus a little note explaining his idea. Jane has got someone with her and simply replies OK to Peter's message. Peter quickly confirms the tickets for the movie and pays for them with his e-wallet that is linked to his PDA device identity (this creates a record that a transaction took place from Peter's PDA, the access to the services themselves are charged, but a small fee is levied from the transaction by Peter's service provider).

Jane finishes her impromptu meeting and quickly looks to her presence service to see if she can tell where Peter is from her buddy list. She can see that Peter is in the middle of town. She quickly calls him as she is leaving the office simply by selecting the icon of Peter in her buddy list. They quickly arrange to meet in a bar where they can decide on dinner. Jane is having second thoughts about the Italian. Having browsed a few more menus together from the bar, they decide on a local Chinese restaurant.

After the meal and movie, they leave the cinema to find their train has been cancelled, no driver again, this was sent to both their PDAs as part of the travel updates service they subscribe to, better find a cab.

When the couple arrive home, both Jane's and Peter's handset swap to DECT for voice and Bluetooth for their home network. Peter and Jane's PDAs now share a common number and are both just handsets of the same DECT base station. When someone calls the house number, both PDAs will be paged. However, since the PDAs are intelligent devices and can receive caller information over the DECT interface they can apply a local profile that can decide whether to ring (playing a distinctive tune for different callers) or not.

Peter and Jane both retire for the evening leaving their PDAs in their cradles.

This simple story highlights some of the capabilities of a combined services next-generation network. All of the service features described will be possible in the next 10 years or less, in fact some are appearing at the time of writing!

Part III: Implications

INTRODUCTION

In this final part of the book, I try to expand on the previous chapters and explore a view of the world of next-generation services. At the time of writing the telecommunications world is undergoing one of its worst downturns in history. The market place is littered with Internet companies going bankrupt and even the big players suffering badly (Lucent's need to find funding and even a risk of having to file for Chapter 11 (protection against bankruptcy), Nortel shedding 15,000 staff) from the downturn of the US market and the knock-on effects around the world.

The bleak picture the author believes is only a short-term hiatus in the market place that will see a renaissance of new profitability for the telecoms operators and suppliers that survive. New services will be aggressively marketed as life-style products that enhance personal productivity and corporate services that will enhance Business-to-Business (B2B), Business-to-Enterprise (B2E) and Business-to-Consumer (B2C) interactions via Application-to-Application (A2A) based communications enabled via XML in the form of SOAP, UDDI and WSDL.

The next 3 years will see major investment (by those that can), divestments by business that are refocusing their strategies and preparation by telecoms operators to move their networks from the current TDM based infrastructures to packet based systems. In the UK Hutchison 3G and Cable&Wireless have already, publicly committed resources and money in third-generation mobile networks and an enhanced packet based backbone networks, respectively. This trend will continue over the next 10 years, it will take that long to transform the networks from the PSTN we now have to the multimedia future with more players both existing and new entrants such as ISPs and ITSPs deriving revenue from voice and data services and competing with conventional telecoms operators in this space.

The largest force of change in the industry now is the declining margin from basic voice services, as they reach commodity status. That is not to say there is no revenue left in voice, just not as much profit. Telcos have embraced the next-generation network infrastructures as a means of transforming their businesses to seek new higher margin products.

Bandwidth has also reached a state of excess; technologies such as DWDM have created a glut of capacity in anticipation of the broadband future. Alas, that future has not transpired as quickly as the original market hyperbole forecast. The upshot is an industry waiting for something to happen. The something is broadband access to eat up all this backbone capacity with revenue generating services. The broadband local loop has been slow to start in all but Singapore, caused by weak regulators and strong incumbents. The UK has seen a number of competitive local access providers suffer because of the combination of the slow process of unbundling and a major decline in expected demand.

Based on this rather painful start to the future telecommunications network, what are Telcos to do? I would like, in this final part to the book, to explore some of the paths telecoms operators, Service Providers and competitive carriers could take (and are taking) to the future and to explore some ideas on how the market is changing and how this affects the suppliers of technologies and services to the telecommunications market place.

14

Expectation and Realisation

14.1 TOO MUCH TOO SOON?

The last decade has seen massive acceleration in the introduction of new technologies and what seemed to be an insatiable demand for IT. Just look at the changes that we have seen: WAP, GPRS, xDSL, DWDM, VoIP, xSPs and hosted services, Bluetooth, B2B and B2C, XML, Java, and others. The thirst for new things has made the market fickle; just look what's happened to Wireless Application Protocol (WAP). WAP was/is basically a good idea with a good foundation just oversold (some might argue under-delivered?). As the hype fell away and the realisation set in that WAP over 2G data (9.6 kbps) wasn't going to set the world on fire, just as readily as the investment came, the industry and media have nailed the coffin shut! What's interesting is, in the Far East NTT DoCoMo's equivalent service (i-mode) has had a resounding success in this area (in excess of 24 million subscribers!).

The free lunch syndrome could well have fuelled this demand. The Internet model seemed to imply you could have services for free and as Internet Service Providers (ISPs) strove to offer free access numbers, this vision became even more believable. The issue was that revenues had to be made somewhere and the search centred on selling advertising space in portals. Voice seemed to be a no go area even companies with a massive investment in voice were treating voice as an evil product and turning their backs on it as quickly as they could to rush to invest in the Internet bubble. The massive investments (and greed) with no return soon created a lack of confidence in the market that unfortunately resulted in collapse of investment and lack of confidence that has not only affected the Internet start-ups, but the solid telecommunications businesses.

Now it's easy to make these observations in hindsight, it always is, but what can be learned from this? The lesson is telecommunications takes time, look back over its long history and you see incredible effort put into getting things right, without the massive investment in the standardisation process, we would have still been using operators or connect international calls, because the different 'phone systems wouldn't have been compatible.

I believe the telecoms industry has now learned this lesson and is concentrating its efforts and finances on structured longer term projects with structured investment and key measurable deliverables, for example Hutchison 3G, in the UK, is backed by Hutchison Whampoa, NTT DoCoMo and KPN. Hutchison 3G are looking to become a major force in third-generation (3G) mobile telecommunications by building a new 3G infrastructure, unencumbered by having to write off investment in existing 2- and 2.5G equipment.

This enforced slowdown is good news for everyone; the previous pace of change could not have been sustained.

14.2 WHERE TO NOW?

The challenge for operators in the telecoms space is clearly to make good of the convergence of the voice and Internet worlds. In order to continue to gain revenue from a product with declining margin (voice) two options present themselves: cut the cost of provision or subsidise the cost from other products. Voice isn't going away, people like to talk.

The options are actually both possible with the technologies available to telecoms network operators. Cost cutting can be made through better, slicker, provisioning of voice services. The World Wide Web (WWW) has presented an excellent opportunity to allow customers to order, provision and control their own services. This combined with open systems, software frameworks and common Application Programming Interfaces (APIs) from initiatives such as Operational Support System (OSS) through Java, Java APIs for Integrated Networks (JAIN), Java 2 Enterprise Edition (J2EE) and .NET should make it possible to reduce the cost of the process of acquiring and providing for new customers and will provide the software necessary to construct the new componentised services. Next-generation network softACD and Customer Relationship Management (CRM) should assist organisations in providing customers with post sales care and through increased use of customer self-help systems, drive down cost in this area too. The key message here is about using the technology to the best end to augment business processes and maintain customer satisfaction.

Subsidisation of dwindling voice margin with more lucrative products may sound like not such good business practice and may even appear to be illegal under some licensing conditions. What I'm not talking about here is the practice of undercutting competition by subsidising a loss making business, but about the use of voice as a key service pull through. We've already stated voice isn't going to go away, so capitalise on it. Use service bundling as a means of attracting revenue from other services. Think about what makes mobile network operators substantial revenue (and it's there part of the basic build, short message services). What's a good fixed line equivalent – voicemail is a great value-added service. Bundling voice and dial-up Internet access together with per-month subscriptions is creating guaranteed revenue for a number of telcos offering that service. Service charges are the profit earners, whilst usage costs will reach the point of pretty much zero return on investment.

Another area of focus is around which of the two orthogonal approaches: specialisation or diversification to take. Specialisation clearly has greater risk, as any downturn in the specialist area can create big problems for a business,[1] but it also carries with it an opportunity to prosper where specific specialisation is in great demand. Diversification can create security from areas of downturn, but can also cause solutions that are being marketed to be too woolly and imprecise. The future telecoms marketplace will support both these strategies and will create a multi-tier multi-facetted market space. Figure 14.1 tries to capture this idea.

Diversification can be created via packaging of other service provider's services and for example, the concept of the virtual Application Service Provider (vASP) can be realised. Nortel have entered the space of service provision by offering to take over telecoms operators' circuit switched networks and run them for a service charge, whilst also introducing new packet voice based equipment as part of a migration process. Specialisations are realised lower down the strata as suppliers of specific service elements, such as voicemail platforms, routers or wholesale (leased) bandwidth providers.

Specialisation can also be represented by companies divesting parts of their business to become more focused on what are considered core competencies, for example Alcatel's recent announcement to outsource its consumer products manufacturing capability. A number of telecoms operators are looking to outsource even their current circuit switched network infrastructures to focus on service provision, in an aim to move up the food chain to higher profit margin products.

---

[1] At the time of writing Marconi announced job cuts of the order of 4000 staff across the UK and the closure of one of their sites. This has been blamed on the downturn, but also on Marconi's decision to specialize in telecommunications by getting out of defence contracts.

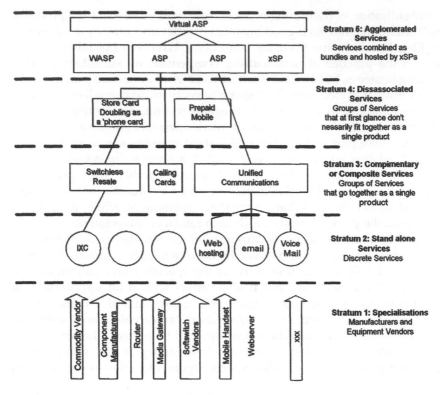

Figure 14.1 Market segmentation specialisation vs. diversification

The infrastructural glue that will allow this service tier-ing is Extensible Markup Language (XML). At each stratum in the model above XML will be used to define the service (Web Service Description Language (WSDL)), interact with the service (Simple Object Access Protocol (SOAP)) and allow dynamic discovery and service brokering to take place (Universal Definition Discovery and Integration (UDDI)), whilst IP Data Records (IPDRs) may well provide billing information (see Chapter 8).

What this market separation also creates is greater opportunity for the service industries that serve the telecoms market space to find new opportunities to help telecoms service providers and network operators manage new relationships with vendors and other application/service providers.

Over the last decade the concept of the carriers' carrier appeared. A larger network operator would sell chunks of bandwidth or call minutes from their international networks to other carriers to provide them with global inter-connects at less than it would cost to build their own.

The larger telcos who are now carriers' carriers are in a strong position to become the future service providers' service provider by providing white-label (generic/customisable) services at strata 2 through strata 5 in Figure 14.1.

14.3 STRATEGIES FOR MAKING IT HAPPEN

This is where I would like to have been able to give you, the reader, the path to that elusive 'killer-app' of services; alas, if anyone knows it, I am sure they're not telling. What I will do is give what I believe as some of the important aspects of next-generation networks for you to consider within your own environment. The paragraphs that follow highlight some of the key points of the book and put them in the context of the business decision to be made.

Do not think class 5 switches (local exchange features) are the only way of delivering class 5 features, think softswitches, and be careful not to re-invent the wheel. Only implement those features that your customers really need. Softswitches offer much greater flexibility for integration of other features such as Unified Communications (UC) and presence to existing Plain Old Telephony Service (POTS) style services.

Look to extend the reach of the network as far out to the customer as you can to create tie-in for service provision. This is a key point for those companies looking to capitalise on market changes such as unbundling in the local loop. If access to the local loop is a business goal, look at perhaps providing an application on customers' premises. This might seem counter to the goal of providing hosted services, but not all aspects of the service need to be hosted.

Make strategic partnerships with key content vendors and service providers – for example Hutchison 3G has cut a deal for the streaming delivery of football matches and a deal with 186k for provision of bandwidth.

Look to your Business Support Systems (BSS) and Operational Support Systems (OSS) and ensure they have the flexibility to accommodate and process XML records both for the mediation of XML usage records, and for the exchange of billing information between organisations. This will help streamline communications between partners and reduce cost.

Create products and services that are as flexible as possible and that are endpoint delivery mechanism independent (i.e. multi-channel capable). UC is a good example of this. The main functional component of a UC solution is the in-box. All the other components of a UC solution are about accessing that in-box, irrespective of the means chosen to reach it, email, and voice browsing or web access.

Look to Megaco as a transition architecture to move your existing Public Switched Telephone Network (PSTN) services to a packet switch network infrastructure, only use it where you believe you can't live without some of those class 5 features that the supplier of your Megaco MGC will support or if you need to transition profitable Intelligent Network (IN) services to a packet switched network (for example you may have invested heavily in Service Control Points (SCPs) and need to write off the equipment). Remember the Media Gateways (MGs) are there only to preserve the circuit switched world, when the circuit switches go, so do the media gateways and signalling gateways to the PSTN. When going down this route, think about the future features the media gateway controller will need to provide and consider how important this flexibility is to the future services. Softswitches may be a better choice if they also integrate with media and signalling gateways. Softswitches generally provide a superset of features above the media gateway controllers that are basically re-compiled cousins of the stored program controller code of the particular vendor's circuit switch.

Look to softswitches and application servers as the means to integrate old and new worlds when you have no existing infrastructure to protect. Softswitches should be able to support all the old and new protocols you need and certainly should be capable of interfacing to and controlling MGs where necessary. Think softswitch for call control, and application servers for serious integrated enhanced services.

If you are a fixed network vendor, remember Fixed Mobile Integration (FMI), we all tried in various ways to make this happen in the circuit switched world. I think its time has come again, look to create strategic partnerships with 3G network operators. As palmtop devices get location capabilities, wouldn't it be a good offer to a customer if they only ever needed one device, whether they were at home or on the move. Create services that can capitalise on the location and route calls based on it (see the example service at the end of Chapter 13).

Look to use directories as much as possible to integrate your Business Systems (BSS). Meta-directories (see Chapter 9) are an important tool for gluing together your business systems with business processes – only change information once and see the power of it being reflected in all your other systems.

Think of customer care as the pinnacle of your organisation's focus towards the customers, if customer care can see all the information necessary to keep your customers happy, they can monitor the customer life cycle. With all the information glued together through a meta-directory, marketing can use this information to target campaigns.

Create services that customers can control through simple interfaces, if it is easy for customers to request services via web pages and have them provisioned automatically then take-up will be quicker and less costly (if you get it right).

Promote mixed media access to services and make good use of Internet Protocol Interactive Voice Response (IP-IVR) to enhance services with voice. Speech recognition is a powerful tool when used with web access and text-to-speech as a voice portal service. Mobile devices don't have a lot of room for keyboards and large displays. Having an intelligent service linked to a diary and speech recognition can make organising your day much easier.

Think about how you can make your services generic, white-label services can be sold wholesale and re-labelled. This has implications on the platforms you use and how they are managed. Think about having to tier your management interfaces to allow different levels of access and partitioning of the view of services. Think about how management interfaces and management information can be customised through the use of XML and style sheets.

As a service industry company, look at how the market segmentation (Figure 14.1) is creating the need for new service level agreements and contracts between equipment vendors' service providers and telcos; and service provider to service provider. And look for consultative service opportunities around process change and technology change. Systems integrators have a lot to offer telcos and xSPs, both parties need to recognise this and make good the services offered.

Take care when constructing the packet-based infrastructure to carry real-time audio signals; do not put cost reduction before quality (see the closing remark in Section 2.2 on voice compression and quality). Care must also be taken on what is carried in the voice encoded bit streams, modem traffic and fax calls don't take too kindly to being exposed to voice compression schemes. Keep track of latency, packet loss and desequencing of packets, all of these will affect the listener quality of the voice. These considerations can be relaxed to some degree if the network is to carry broadcast traffic such as near video on demand, as buffering of the transmitted signals can be employed (at the expense of increased latency).

Don't treat all voice gateways as the same just because they provide the same encoding algorithms, the techniques to cope with packet loss, jitter and timing when converting between circuit switched telephony and packet switched telephony are at the time of writing proprietary to the vendor of the gateway.

Figure 14.2 shows the elements that fit together to form the business of next-generation networks, think about how these fit together to provide services you can sell.

The main commodity components for new services (as indicated in Figure 14.2) are content, bandwidth, mobility and fixed access. The combination of these into new services and the delivery model of these services are where the service providers come in.

Customers will be looking to service providers to provide the complete life cycle of products, from delivery through to help and maintenance. The next-generation network services will consist of complex components that all need to work together in a seamless way. In order to market these new services as life-style products, not only do the products have to be mass market, but they also have to be easy to maintain and cheap to maintain. Nokia have taught everyone a valuable lesson – mobile phones can be fashion statements, their express-on™ covers are extremely cheap to manufacture but sell at a high margin.

Figure 14.2 Business mix

14.4   HOW LONG AND HOW MUCH?

Clearly, the title for this chapter is the '$64,000,000 question'. The changes that will make the next-generation networks a reality are some time off yet, but don't dismay some services will start to appear in the next 2–3 years. The kind of service scenario described at the end of Part 2 is probably at the worst case 10 years away. Table 14.1 highlights the developments that will likely take place as the shape of the next-generation networks starts to form.

What will make this happen are changes to the way companies do business. Companies with a traditional product or service approach, currently compete for the last morsel out of a market until a market reaches commodity status and there is no value left. The next-generation business will have to work in a more collaborative way, passing on revenue through a sequence of business-to-business transactions, some of which might individually only account for cents worth of revenue, but because of the sheer volume of transactions the actual revenue amounts to much more. This is a micro-payment model, where each business in the chain to delivering a service to a consumer of that service, levies a little from each transaction.

In this new approach to the market, the competition will be over who can create the most sought after bundles of services. The deals each

Table 14.1 Next-generation network timeline[a]

2000	2001	2003+	2010
Starting position	Desired future state starts to take shape	Major installation and investment programmes underway. Mergers and acquisitions continue	All major telcos around the world have integrated voice and data on to a common infrastructure. FMI is complete in some countries and global roaming is available. Some Third World nations are still struggling to keep up
Networks fragmented, voice on one network, data on another	First commercial launch of GPRS happens	Early adopter next-generation networks are offering first telco grade products. Issues with service interaction and protocol inter-working largely fixed	Content is more important than how it got there. Things like dialling a number is obsolete, why dial when you can speak 'Call John'
IP becoming the predominant protocol	Standards based IP on telco grade systems starts to appear from equipment manufacturers	Business-to-business products such as VPNs and IP Centrex launched and a number of major call centres are using softswitches to deliver network CTI to a distributed call centre environment	Home network integrated into a broadband local loop, new homes are built with wireless networking entertainment in mind

Table 14.1 (*continued*)

2000	2001	2003+	2010
ITU-T, ANSI and ETSI signalling prominent in voice networks	IP billing reaches commercial viability for differentiated billing based on service offered. New standards emerge based on XML (IPDR)		Global commitment to standards and regulation ensure interoperability issues are fixed
Voice networks congested with ISP dial-up traffic			Moore's law finally gives out, limits of semiconductor technologies are reached and optical networks become prevalent for distribution of ultra wideband services
VoIP standards gathering speed (MEGACO, MGCP and SIP)		Wireless local loop sees a resurgence as a means of delivering service. Mobile devices are location sensitive and adopt a location specific service	
VoIP established in the enterprise using H.323	Trials of UMTS services are in full swing in the mobile marketplace and early releases take place in Japan	FMI starts to take shape again, after a shaky start in the mid-1990s	The promise of e-commerce is finally realised, as home shopping takes on a new meaning. Applications and games are provided electronically (on loan). Video on demand allows Blockbuster 2000 to distribute the latest films. Entertainment companies can no longer restrict full distribution to different countries at different times

Table 14.1 (continued)

2000	2001	2003+	2010
Telcos trialing IP for core network transport of voice. Looking for cost savings			
Telcos and service providers looking for next-generation network equipment suppliers	ISPs and telcos consolidate to avoid broadcasters taking over content market	Europe becomes a major player in the global marketplace and GSM/UMTS drives business communications across continental Europe	

[a] FMI = Fixed Mobile Integration; GPRS = General Packet Radio Service; VPN = Virtual Private Network; IP = Internet Protocol; CTI = Computer Telephony Integration; ITU-T = International Telecommunications Union telecommunications; ANSI = American National Standards Institute; ETSI = Europe Telecommunications Standards Institute; ISP = Internet Service Provider; XML = Extensible Markup Language; IPDR = IP Data Record; UMTS = Universal Mobile Telecommunications Service; GSM = Global System for Mobile Communications.

vendor of a micro-service can make with other vendors wanting to bundle that micro-service will be what drives the market and will create a greater opportunity for entrepreneurial innovative services.

The investments to make this happen will have to be in more than just infrastructure. Organisation will have to invest in transforming their business processes to accommodate better integration at all layers from customer acquisition through provisioning and fault management to finally billing. Greater numbers of inter- and intra-company transactions will be performed electronically and automatically enhanced by XML documents. How much are businesses going to have to invest I can't say, but based on predictions from organisations like Ovum, this change will save organisations money in the longer term.

What I can say however is that with any major change in both technology innovation and social behaviour, this change comes at a price. The price is generally more than commercial; it involves a whole change in the corporate emotional state. Businesses at the end of the day are about people and the way they interact, telecommunications is about making that communication easier, however, every once in a while a new technology or way of doing things comes along that requires a step change in behaviour for that change to take place. Stephen Covey is famous for his Seven Habits of highly effective people, he highlights in this book the emotional energy and investment involved in undertaking what Thomas Kuhn coined 'a paradigm shift'. We (the telecoms industry) are having to go through this pain of a paradigm shift in the telecommunications industry, from circuit switched based voice networks to packet-based integrated networks.

What has epitomised the telecoms business of the last decade is the speed at which everyone has had to accept change. The latest change to a fully integrated network has been evolving over some time, but we are now at what Geoffrey A. Moore refers to as the 'chasm', the gap between the early adopter and the early majority. Voice over IP has been around in academia and in live trials for a few years, but has really only gained any subscription amongst the early adopters and innovators, to make the opportunity a reality in Geoff Moore's words "we have to cross the chasm".

I hope we make it!

References and Further Reading

[BLACK1] Black, Uyless & Waters, Sharleen, SONET T1: Architectures for Digital Transport Networks, Prentice-Hall Series in Advanced Communications Technologies, Prentice-Hall, Upper Saddle River, NJ.

[BLACK2] Black, Uyless, ATM Volumes 1, 2 & 3 – Foundation for Broadband Networks, Prentice-Hall Series in Advanced Communications Technologies, Prentice-Hall, Upper Saddle River, NJ.

[BLACK3] Black, Uyless, Mobile and Wireless Networks, Prentice-Hall, Upper Saddle River, NJ.

[BRAD] Bradner, Scott & Mankin, Alison, IPng Internet Protocol Next Generation, Addison-Wesley, Reading, MA.

[BTTECH] O'Neill, AW & Tsirtsis, G, Edge mobility architecture – routing and handoff, BT Technology Journal, Vol. 19, No. 1, January 2001.

[CHRIS] Christensen, Gerry, et al., Wireless Intelligent Networks, Artech House, Norwood, MA.

[CPIM] A common profile for instant messaging, IETF internet draft, draft-ietf-impp-cpim-01.

[DAVI] Davie, Bruce & Rekhter, Yakov, MPLS Technology and Applications, Morgan Kaufmann, Orlando, FL.

[DYNA] Bennett, Russell & Rosenberg, Jonathan, White paper – Integrating presence with multimedia communications, Dynamicsoft.

[DYSA] McDysan, David E & Spohn, Darren L, ATM Theory and Application, McGraw-Hill, New York.

[EBERS] Eberspacher, Jorg, et al., GSM Switching, Services and Protocols, John Wiley, Chichester.

[FAYN] Faynberg, Igor, et al., The Intelligent Network Standards – Their Application to Services, McGraw-Hill, New York.

[HALS] Halsall, Fred, Data Communications, Computer Networks and Open Systems, 4th edn, Addison-Wesley, Reading, MA.

[HERS] Hersent, Olivier, et al., IP Telephony – Packet-based Multimedia Communications Systems, Addison-Wesley, Reading, MA.

[HUIT] Huitema, Christian, Ipv6 the New Internet Protocol, Prentice-Hall, Upper Saddle River, NJ.

[IAKO] Venieris, Iakovos & Hussmann, Heinrich (Eds), Intelligent Broadband Networks, John Wiley, Chichester.

[JOHN] Johnson, Alan B, SIP – Understanding the Session Initiation Protocol, Artech House, Norwood, MA.

[KAMP] Kampman, Kevin & Kampman, Christina, All about Network Directories – Understanding Directory Services and Business Applications, John Wiley, Chichester.

[LES] Voice and Multimedia Over ATM–Loop Emulation Service Using AAL2, AF-VMOA-0145.000, ATM Forum.

[MANN] Mann, Steve, Programming Applications with the Wireless Application Protocol, John Wiley, Chichester.

[MARG] Margulies, Ed, The UnPBX – The Complete Guide to the New Breed of Communications Servers, Flatiron.

[ORFA] Orfali, Robert, et al., The Essential Distributed Objects Survival Guide, John Wiley, Chichester.

[OVUM] Softswitches: the Keys to the Next-generation IP Network Opportunity.

[PERS] Lindgren, Pers, A multi-channel network architecture based on fast circuit switching, PhD thesis.

[RFC2778] A Model for Presence and Instant Messaging IETF RFC 2778.

[RFC2976] SIP INFO Message.

[RFC3050] Common Gateway Interface for SIP.

[RUSS] Russell, Travis, Signaling System #7, 2nd edn, McGraw-Hill, New York.

[SCRIB] Scribner, Kennard & Stiver, Mark C, Understanding SOAP, SAMS.

[SIEG] Redl, Siegmund M, et al., GSM and Personal Communications Handbook, Artech House, Norwood, MA.

[SILL] Siller Jr, Curtis A & Shafi, Mansoor, Synchronous optical network synchronous digital hierarchy: an overview of synchronous networking, SONET and SDH – A Source Book of Synchronous Networking, IEEE Press.

[SOLO] Solomon, James D, Mobile IP the Internet Unplugged, Prentice-Hall International.

[STALL] Stallings, William, ISDN and Broadband ISDN Frame Relay and ATM, Prentice-Hall International.

[STEV] Stevens, W Richard, TCP/IP Illustrated, Vols 1–3, Addison-Wesley, Reading, MA.

[TANE] Tanenbaum, Andrew S, Computer Networks, 3rd edn, Prentice-Hall, Upper Saddle River, NJ.

[WAPF] WAP Forum Ltd, Official Wireless Application Protocol: The Complete Standard with Searchable CD-ROM, John Wiley, Chichester.

[WARR] Warrier, Padanmand & Kumar, Balaji, XDSL Architecture, McGraw-Hill, New York.

[WHITEH] http://www.whitehouse.gov/history/presidents/rh19.html.

[WRIG] Wright, David J, Voice over Packet Networks, John Wiley, Chichester.

Glossary

A2A	Application to Application
AAL	ATM Adaptation Layer
ACD	Automatic Call Distributor
ANSI	American National Standards Institute
ARP	Address Resolution Protocol
ASP	Application Service Provider
ATM	Asynchronous transfer mode
B2B	Business to Business
B2C	Business to Consumer
BCSM	Basic Call State Model
Bluetooth	Bluetooth
BSS	Base Station System
BSS	Business Support System
CAMEL	Customised Application Mobile Enhanced Logic
CAP	CAMEL Application Part
CAS	Channel Associated Signaling
CDMA	Code Division Multiple Access
CGI	Common Gateway Interface
CIM	Common Information Model
CS	Call Server
CPE	Customer Premise Equipment
CPL	Call Processing Language
CRM	Customer Relationship Management
CSS	Cascading Stylesheet
CSTA	Computer Supported Telephony Applications
CTI	Computer Telephony Integration
DECT	Digital European Cordless Telephony
DEN	Directory Enabled Network
DWDM	Dense Wave Division Multiplexing
DNS	Domain Name System
DSL	Digital Subscriber Line
DSP	Digital Signal Processing
DTM	Dynamic Synchronous Transfer Mode
DTMF	Dual Tone Multifrequency

ECTF	Enterprise Computer Telephony Forum
EDI	Electronic Data Interchange
EDGE	Enhanced Data GSM Evolution
ETSI	Europe Telecommunications Standards Institure
FDM	Frequency Division Multiplex
FMI	Fixed Mobile Integration
FPLMTS	Future Public Land Mobile Telephony Service
3GPP	Third Generation Partnership Programme
GPRS	General Packet Radio Service
GTT	Global Title Translation
HTML	Hypertext Markup Language
HSCSD	High Speed Circuit Switched Data
IAD	Integrated Access Device
IETF	Internet Engineering Task Force
IN	Intelligent Network
INAP	Intelligent Network Application Protocol
IP	Internet Protocol
IPDR	IP Data Record
ISDN	Integrated Services Digital Network
ISP	Internet Service Provider
ISUP	ISDN User Part
IVR	Interactive Voice Response
JAIN	Java APIs for Integrated Networks
LDAP	Lightweight Directory Access Protocol
LAN	Local Area Network
LNP	Local Number Portability
MAN	Metropolitan Area Network
MEGACO	Media Gateway Control
MGCP	Media Gateway Control Protocol
MPLS	Multi Protocol Label Switching
MVNO	Mobile Virtual Network Operator
NAT	Network Address Translation
OC	Optical Carrier
OSPF	Open Shortest Path First
OSS	Operational Support System
PAM	Pulse Amplitude Modulation
PANS	Pretty Amazing New Services
PBX	Private Branch Exchange
PCM	Pulse Code Modulation
PDA	Personal Digital Assistant
PDH	Plesiochronous Digital Hierarchy
PLMN	Public Land Mobile Network
POP	Point of Presence
POTS	Plain Old Telephony Service
PPP	Point to Point Protocol
PPTP	Point to Point Tunneling Protocol

PRI	Primary Rate Interface
PSTN	Public Switched Telephone Network
RTP	Real Time Transport Protocol
RTCP	Real Time Stream Control Protocol
RTSP	Real Time Stream Protocol
SCCP	Session Connection Control Part
SCE	Service Creation Environment
SCTP	Stream Control Transmission Protocol
SDH	Synchronous Digital Hierarchy
SDP	Service Data Point
SIB	Service Independent Building Block
SIGTRAN	Signalling Transport Group
SIP	Session Initiation Protocol
SLP	Service Logic Program
SME	Small to Medium Business
SNMP	Simple Network Management Protocol
SOAP	Simple Object Access Protocol
softswitch	Softswitch
SONET	Synchronous Optical Network
SP	Service Provider
SPC	Stored Program Controller
SS#7	Signalling System Number 7
SSP	Service Switch Point
Strowger	Strowger Exchange
STM	Synchronous Transport Module
STP	Signalling Transfer Point
STS	Synchronous Transport Signal
TAP	Transfer Account Procedures
TC	Transmission Convergence layer
TCAP	Transaction Capabilities Part
TCP	Transport Control Protocol
TDM	Time Division Multiplex
TMN	Telecoms Managed Network
TUP	Telephony User Part
UDDI	Universal Definition Discovery and Integration
UDP	User Datagram Protocol
UM	Unified Communications
UMTS	Universal Mobile Telecommunications Service
USB	Universal Serial Bus
VC	Virtual Channel
VCI	Virtual Channel Identifier
VoiceXML	Voice XML
VoIP	Voice over IP
VPI	Virtual Path Identifier
WASP	Wireless Application Service Provider
WAP	Wireless Access Protocol

WDM	Wave Division Multiplexing
WML	Wireless Markup Language
WMScript	Wireless Markup Script
WSDL	Web Service Description Language
W3C	World Wide Web Consortium
XHTML	Extensible Hypertext Markup Language
XML	Extensible Markup Language
VRU	Voice Response Unit
XSL	Extensible Style Language
VPN	Virtual Private Network
WAN	Wide Area Network

Index

Printed and bound by CPI Group (UK) Ltd, Croydon, CR0 4YY

27/10/2024

14580206-0005